Contents

KU-492-143

Preface

Who is this book meant for? It is for anyone who is interested in oscilloscopes, how to use them and how they work, and for anyone who might be if he or she knew a little more about them.

It is easy to say what the book is not: it is not a textbook of any sort, and particularly not a textbook on how to design oscilloscopes. Nevertheless, besides describing a great variety of oscilloscopes, their particular advantages and how to use them, the book explains briefly how these instruments work, on the basis that the best drivers have at least some idea of what goes on under the bonnet. This takes us into electron physics and circuit theory—but not too far. Formulae and results are simply stated, not derived or proved, and those with only the haziest knowledge of mathematics will find nothing to alarm them in this book. Consequently, readers in their earliest teens will be able to learn a lot from it; Chapter 1 is written especially for anyone with no prior knowledge of the subject. Sixth-formers and students on ONC and HNC courses should all find the book useful. Even many degree students will find it of considerable help (though they may choose to skip Chapter 1!); electronic engineering undergraduates have plenty of opportunity to learn about oscilloscopes, but many graduates come into electronic engineering from a physics degree course, and will welcome a practical introduction to oscilloscope techniques.

Technicians and technician engineers in the electronics field will of course be used to oscilloscopes, but the following chapters should enlarge their understanding and enable them to use the facilities of an oscilloscope to the full. Finally, I hope that those whose interest in electronics is as a hobby, including many amateur radio hams and radio-controlled-model enthusiasts, will find the book valuable, especially if they are considering buying or even constructing their own oscilloscopes.

I.H.
1981

Oscilloscopes

How to use them, how they work

Second Edition

Ian Hickman
BSc (Hons), CEng, MIEE, MIEEE

Heinemann: London

William Heinemann Ltd

LONDON   MELBOURNE   JOHANNESBURG   AUCKLAND

First published by Newnes Books 1981
Reprinted 1984
Second edition 1986
Revised edition
first published by William Heinemann Ltd 1987

ISBN 0 434 907383

Printed in Great Britain by
Thomson Litho Ltd, East Kilbride

SB 40340 £6.95. 489

Preface to the revised second edition

The ability to investigate circuit behaviour directly, to see precisely what is going on, is the most powerful development and servicing tool in electronics and associated disciplines. Little wonder, then, that the pace of oscilloscope development itself shows no sign of slackening. Since the appearance of the second edition in early 1986, some of the models there featured have been withdrawn, and replaced by later, more powerful oscilloscopes. This revised second edition reflects these changes, all the models illustrated being current at the time of writing, many only recently introduced. The author has also taken the opportunity to add further material to Chapters 4, 5 and 6 and make other minor improvements and corrections, whilst Appendices 1 and 2 have both been revised and updated.

I.H.
January 1987

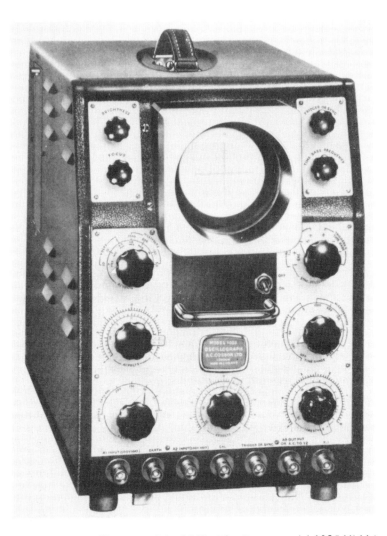

An advanced oscilloscope of the 1940s. The Cossor model 1035 Mk11A was a true dual beam oscilloscope with a maximum bandwidth of 7 MHz (Y1 amplifier), 100 kHz (Y2 amplier) and a fastest sweep rate of 15 μs per scan, with repetitive, triggered and single-stroke operation (courtesy Cossor Electronics Ltd)

1

Introduction

The cathode-ray oscilloscope is an instrument designed to display the voltage variations, periodic or otherwise, that are met with in electronic circuits and elsewhere.

The word is an etymological hybrid. The first part derives from the Latin *oscillare*, to swing backwards and forwards; this in turn is from *oscillum*, a little mask of Bacchus hung from the trees, especially in vineyards, and thus easily moved by the wind. The second part comes from the Classical Greek *skopein*, to observe, aim at, examine, from which developed the Latin ending *-scopium*, which has been used to form names for instruments that enable the eye or the ear to make observations. For some reason the subject of the design and use of oscilloscopes is generally called not oscilloscopy but oscillography, from oscillo- and *graphein*, to write.

There are other types of oscilloscope besides those using cathode-ray tubes. For example, pen recorders, ultra-violet chart recorders and XY plotters are oscilloscopes or oscillographs of a sort, as indeed is 'Fletcher's Trolley' of school physics fame. However, this book is concerned mainly with cathode-ray oscilloscopes.

Representing a varying voltage

The basic principle of oscillography is the representation, by graphical means, of a voltage that is varying. The voltage is plotted or traced out in two-dimensional 'Cartesian' coordinates, named after Descartes, the famous French seventeenth-century philosopher and mathematician.

Figure 1.1 shows the general scheme for the representation of any two related variables. Both positive and negative values of each variable

1

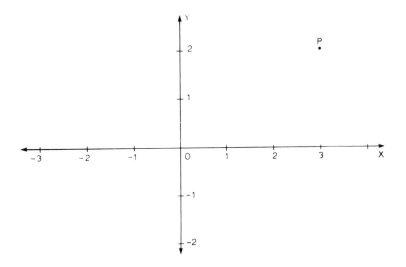

Figure 1.1 Cartesian or graphical coordinates. The horizontal and vertical axes may be to different scales, even different units, for graphical purposes

can be represented. The vertical axis is called the Y axis, and the horizontal the X axis. The point where the axes cross, i.e. where both X = 0 and Y = 0, is called the 'origin'.

Any point is defined by its X and Y coordinates. Thus the point marked 'P' in the top right-hand quadrant is the point (3, 2), because its distance to the right (called its 'abscissa' or X coordinate) is 3 units and its distance up (called its 'ordinate' or Y coordinate) is 2 units.

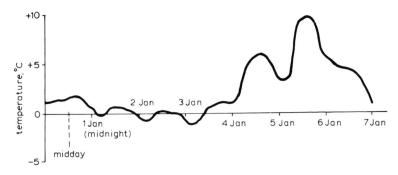

Figure 1.2 Fictional plot of temperature in first week of January. An example of a graph where the horizontal and vertical axes are to different scales and in different units

Figure 1.2 is an example of a graph plotted on Cartesian coordinates, and shows an imaginary plot of the temperature during the first week of January. Quantities that vary with time, like temperature and voltage, are very important in engineering and are frequently represented in graphical form. As we don't usually attribute much meaning to the concept of negative time, the Y axis (the vertical line corresponding to the point where X = 0 or the start of 1 January) has been shown at the extreme left. The X axis now represents time, shown in this case in days, though for other purposes it might be minutes, seconds, or microseconds (usually written μs and meaning millionths of a second). Negative temperatures are plotted below the X axis, and positive temperatures above it. Time is taken as increasing (getting later) from left to right, starting at zero at the origin. Thus the X axis is a 'timebase', above and below which the related variable (in this case, temperature) is plotted.

Voltages can be positive or negative, just like temperatures. The usual reference point for voltages is taken as earth or ground. This is called zero volts, 0V, just as $0°C$, the melting point of ice, is taken as the reference for temperatures.

What the oscilloscope shows

Where you or I might draw a graph like *Figure 1.2* with a pencil, an oscilloscope draws its 'trace' with a moving spot of light on the screen

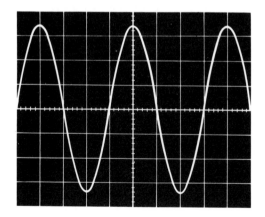

Figure 1.3   240 V a.c. mains waveform, displayed at 100 volts per division vertically and 5 milliseconds per division horizontally.

3

of a cathode-ray tube. The screen is approximately flat and coated on the inside with a powder that emits light where it is struck by a beam of electrons. More about the operation of the cathode-ray tube can be found in Chapter 7; here it is sufficient to note that the internal circuitry of the oscilloscope causes the spot of light to travel from left to right across the 'screen' of the tube at a steady rate, until on reaching the right-hand side it returns rapidly to the left ready to start another traverse, usually called a 'sweep' or 'scan'.

Figure 1.3 shows the picture that might appear on the screen of an oscilloscope if it were used to display the waveform of the 240 V a.c. (alternating current) domestic mains electricity supply. This actually varies between plus and minus 340 V, with a rounded waveform closely approximating the shape known as a sine wave — a very important waveform in electrical engineering. As its positive and negative loops are the same size and shape, the sine wave's 'mean' or average value is zero. The mains is described as 240 V a.c. because that is its 'effective' value; that is to say an electric fire would give out the same heat if connected to 240 V d.c. (direct current) mains, as it does on '240 V' a.c. mains.

The screen of an oscilloscope is often equipped with vertical and horizontal rulings called a 'graticule'. In *Figure 1.3* the scan or X deflection speed corresponds to 5 millisecond per division (5 ms/div). Likewise, in the vertical or Y direction, the sensitivity or 'deflection factor' is 100 volts per division. On oscilloscopes with a 13 cm (5 inch) nominal screen diameter, the divisions are centimetre squares. However, many oscilloscopes have a smaller screen than this. In such cases graticules with fewer centimetre square divisions are sometimes found, but more often smaller divisions are used, to enable the convenient 10 × 8 or 10 × 6 division format to be retained.

'Trigger' circuitry in the oscilloscope ensures that the trace shown always starts at the same point on the waveform. In our example the trace starts as the 240 V a.c. mains voltage is passing through zero, going positive. The frequency of the mains is 50 Hz (Hz is short for hertz and means 'cycles per second'); thus it takes 20 ms to complete each cycle. As the full ten squares of the graticule represent 50 ms in the horizontal direction, two and a half complete cycles are traced out as the spot scans across the screen. During the next half-cycle, the spot returns rapidly to the left of the screen. This return journey is called the flyback, but no trace of it is seen, as the spot is suppressed by a 'flyback blanking' circuit.

The next trace thus commences three cycles after the start of the previous one, so $16\frac{2}{3}$ identical traces are drawn every second. This is not fast enough for the eye to see a single steady picture, so there is a pronounced flicker (unless the cathode-ray tube is a 'long-persistence' type, as described in Chapter 6). If the scan or sweep rate were changed

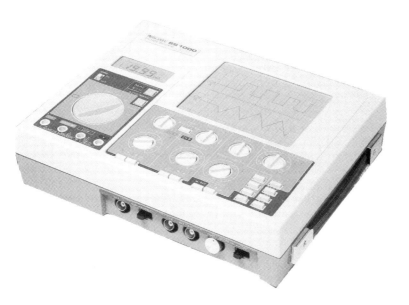

Figure 1.4 The unusual SOAR model 1000 battery-operated digital storage oscilloscope uses a liquid crystal display instead of a cathode ray tube (courtesy House of Instruments (Anglia) Ltd)

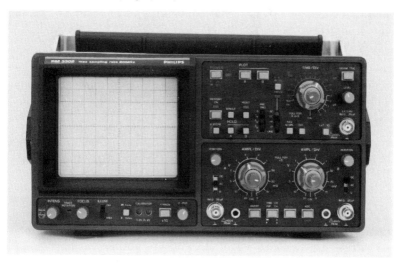

Figure 1.5 The Philips PM3302 offers both real-time operation and digital storage. In the latter mode, events occurring before a trigger signal can be captured (courtesy Pye Unicam Ltd)

5

from 5 ms/div to 20 ms/div, ten complete cycles would appear per scan and the moving spot of light would be seen bobbing up and down as it crossed the screen. On the other hand, if a 500 Hz waveform were viewed at 0.5 ms/div (the same as 500 μs/div), there would be 166 identical traces per second and a completely flicker-free picture would result. However, this is only because the waveform itself is 'periodic', i.e. it repeats exactly from cycle to cycle.

An example of a much more complex waveform that does not repeat exactly is the output of a microphone recording a piece of music. Here, we could never trigger an oscilloscope to give a steady picture, as the waveform itself is constantly changing. The basic oscilloscope, then, is primarily of use for viewing periodic (repetitive) waveforms, although it is often necessary to view single, non-repetitive waveforms; the more expensive oscilloscopes will take this job in their stride also.

Having learnt a little of what an oscilloscope is and what it can do, in Chapter 2 we look in more detail at the facilities provided by a basic oscilloscope.

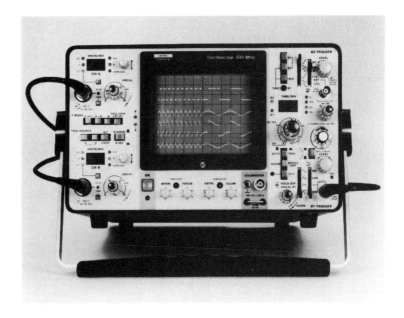

Figure 1.6 One of the few models with a real-time bandwidth of 500 MHz, the 5229 provides digital readout of scale factors, high writing speed and advanced **triggering** facilities (courtesy Solartron Instruments)

2

The basic oscilloscope

Chapter 1 briefly described how an oscilloscope draws its trace with a spot of light (produced by a deflectable beam of electrons) moving across the screen of its c.r.t. (cathode-ray tube). At its most basic,

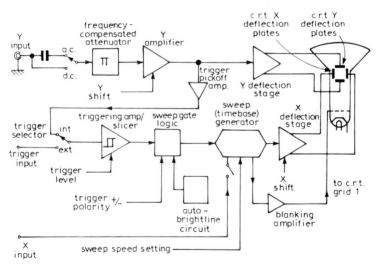

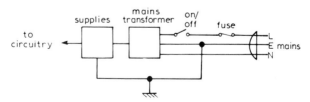

Figure 2.1 Block diagram of basic oscilloscope. Note: It is now common to fit a two pole mains ON/OFF switch, both for safety reasons and to comply with national electrical equipment regulations.

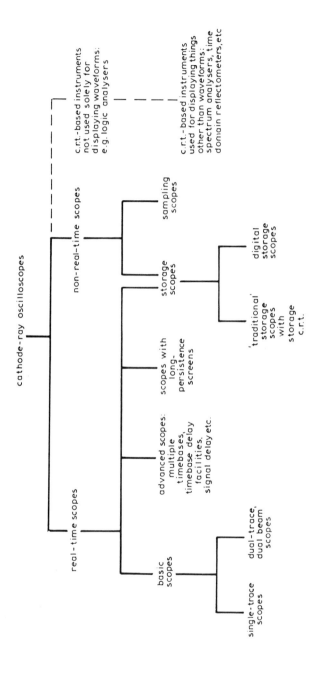

Figure 2.2 Types of cathode-ray oscilloscope

therefore, an oscilloscope consists of the c.r.t. (further details of which can be found in Chapter 7), a 'timebase' circuit to move the spot steadily from left to right across the screen at the appropriate time and speed, and some means (usually a 'Y' deflection amplifier) of enabling the signal we wish to examine to deflect the spot in the vertical or Y direction. In addition, of course, there are a few further humble essentials like power supplies to run the c.r.t. and circuitry, a case to keep it

Figure 2.3 Farnell DT12–5 compact dual-trace 12 MHz oscilloscope (courtesy Farnell Instruments Ltd)

all together, and a Y input socket plus a few controls on the front panel. *Figure 2.1* is a block diagram of such an instrument.

This type of oscilloscope, more or less sophisticated as the case may be, belongs to what is by far the commonest and most important category; it is a 'real-time' oscilloscope. This means simply that the vertical deflection of the spot on the screen at any instant is determined by the Y input voltage at that instant. Not all oscilloscopes are real-time instruments: *Figure 2.2* attempts to categorise the various types available. The distinction between real-time instruments and others is not an absolute and clear-cut boundary, but the fine distinctions need not worry us here.

A really basic oscilloscope then is one with the necessary facilities for examining a repetitive waveform. An instrument with but a single Y input, corresponding to Figure 2.1 and the extreme left-hand branch of Figure 2.2, meets this description. With such an instrument, the relative timing between the waveforms at different points in a circuit can only be established indirectly, by using the external trigger input and viewing the waveforms one after the other. The advantage of being able to see relative timing directly by viewing two waveforms simultaneously is so great that, increasingly, even inexpensive basic oscilloscopes offer this facility. Such an instrument (entirely British designed and built) is shown in Figure 2.3. Although to some readers the facilities it provides may seem entirely self-explanatory, they are in fact worth a closer look.

Basic oscilloscope controls

The mains supply to the Farnell model DT12-5 (see Figure 2.3) was originally controlled by a switch forming part of the 'intensity' or 'brilliance' control, but the instrument is now fitted with a separate ON/OFF switch, a much more convenient arrangement. There is also a mains indicator light. Of course, a light is not usually needed as a warning that one has left the oscilloscope switched on; after all, the trace on the screen does that quite effectively. The indicator's main function is to assure the user that, on plugging in and switching on, the mains socket is live and hence the oscilloscope will be operational as soon as the c.r.t. has warmed up. The traditional British colour for mains indicators is red. The DT12-5 originally had a red mains indicator but the manufacturer has changed this to green. Green is the normal colour for mains indicators in Germany and much of the rest of Europe, red being reserved under IEC regulations for warning lamps. Some manufacturers use orange but the distinction between this and the traditional red can be quite marginal.

The intensity control should normally be used at the lowest setting that gives an adequately bright trace. In particular, if the external X input is selected and no X and Y signals are applied the spot will remain stationary; if the intensity control were then left at maximum for a long period, permanent damage to the screen could occur in the form of a 'burn mark' (an area of reduced screen sensitivity). On the other hand, if examining in detail say a $10\,\mu s$ long pulse occurring once every $200\,\mu s$, it would be necessary to advance the intensity control. This is because, with a suitable timebase setting such as $2\,\mu s/division$, the spot would spend only a tenth of the time writing the trace, and the rest of the time waiting to trigger from the next pulse. But it will be found that, on advancing the intensity control, the trace becomes not only brighter, but thicker. This coarsening of the trace can be largely corrected by adjustment of the focus control, the optimum setting for which depends therefore to some extent on the setting of the intensity control.

There is a limit to how much the intensity can be increased to compensate for low repetition rate of the trace. For example, in the case mentioned above, if the 10 μs pulse occurred once every 100 ms it would not be possible to examine it in any detail on a basic instrument such as that shown in *Figure 2.3*. One would require an instrument with a much higher 'writing speed', a concept more fully explained in later chapters.

Next to the intensity control and centrally placed beneath the c.r.t. screen is the focus control, being the central knob of the group of three. This control should be adjusted to provide the smallest spot size, resulting in the sharpest possible trace. It may need readjustment when viewing low repetition rate waveforms, as explained above. To the right of the focus control is the 'horizontal shift' control, labelled '↔'. There are also 'vertical shift' controls, labelled '↕', one for each input channel. The shift controls enable the traces to be centred horizontally and adjusted vertically so that, for example, zero Y input voltage corresponds to the centre horizontal line. This can conveniently be done with both the input coupling push buttons in the 'in' position, as the Y amplifier is then disconnected from the input socket and grounded. This saves disconnecting the signal being examined from the input socket. The two input coupling buttons provide a choice of a.c. or d.c. coupling. With both released, the corresponding Y trace is not displayed.

For examining voltage variations as a function of time – the main purpose of any oscilloscope – the user must select a suitable timebase speed with the 'time/div' switch. On the DT12-5, speeds of 0.5 μs/div to 500 ms/div in a 1, 2, 5 sequence are provided. Pulling the horizontal shift knob forward provides a × 5 magnification of the sweep, extending the fastest sweep speed to 100 ns/div. One would normally select a speed that results in between two and three complete cycles of the waveform being displayed. Too slow a timebase speed results in so many cycles being displayed that the detailed shape of each cannot be distinguished; too fast a speed results in the display of only a part of one cycle. Likewise, a suitable setting of the volts/div (Y sensitivity) switch, with a.c. or d.c. coupling, should be chosen as required, so that the waveform occupies somewhere between half and full screen height. On early models of the DT12-5 the input attenuator was marked 'volts/cm', but volts/div is more appropriate, even though the graticule divisions are in fact centimetres. Volts/cm is the c.g.s. unit of electric field strength and this is not what an oscilloscope measures.

The volts/div switches provide sensitivities of 5 mV/div to 10 V/div in a 1, 2, 5 sequence. As with the time/div switch, intermediate settings are not available. However, fine sweep and gain controls are provided on the next model up in the range, the DT12-14, together with channel 2 invert and add modes.

The last control function to be mentioned is in many ways the most important: triggering. This topic looms large in later chapters, but on the DT12–5 it is very simply handled by one knob plus five push buttons. The knob's setting controls the point on the input waveform at which the trace triggers. The top push button provides a choice of 'auto' (out) and 'normal' (in) triggering. In the normal mode, the time-base only triggers when a signal is present at the selected signal source (channel 1, channel 2 or external), whereas auto provides the usual 'brightline' in the absence of a trigger signal. The middle button provides a choice of trigger polarity; with the button in, the timebase triggers on negative-going edges. Pressing one or other of the bottom two buttons selects channel 1 or channel 2 as the trigger source, as appropriate. If neither is pressed, then the EXTernal trigger input is selected. The waveform from which the trigger circuit works will be the same as that displayed on the screen when internal triggering (channel 1 or 2) is selected. When using external trigger, the waveform at the external trigger input socket may be of a different *shape* from the trace on the screen but it should generally be of the same *frequency*, or it will not be possible to obtain a stable, locked display. It may alternatively be a submultiple of the frequency of the displayed waveform, or (more difficult to use successfully) a harmonic of it.

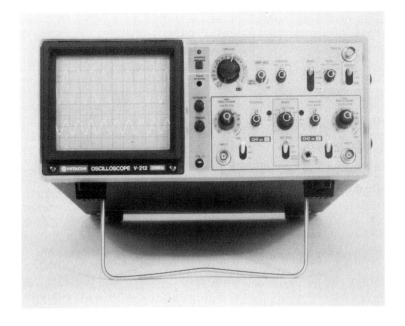

Figure 2.4 The Hitachi V212 budget-priced dual trace 20 MHz portable oscilloscope has a large screen with internal graticule and 1 mV/division sensitivity (courtesy Hitachi Denshi (UK) Ltd)

For certain purposes it may be desired to deflect the spot in the X direction not from the oscilloscope's internal timebase generator, but from some external waveform. This may be achieved by selecting the X–Y position (fully anticlockwise) of the time/div switch and connecting the waveform to the channel 1 input, which doubles as the X input.

Unusually for a budget priced oscilloscope, the DT12-5 includes a choice of a.c. or t.v. triggering. The latter is fully automatic from normal or inverted composite video signals. An internal sync separator triggers the scope from the field sync pulses unless the timebase speed selected is 50 μs/div or faster, in which case line sync is used. The time/div switch also controls the selection of 'chopped' or 'alternate' display. The chop mode is used when displaying both Y inputs with time/div set to 1 ms/div or slower. The chopped mode of display is described further in Chapter 3. Above 1 ms/div the two traces are displayed alternately. The final front panel item is a socket providing a 1 V ±2 per cent peak-to-peak squarewave at 1 kHz. In addition to its use in adjusting probe compensation (see Chapter 4), this may be used to check the calibration of the vertical channels.

The rear panel carries an input socket for Z modulation, and screwdriver operated preset controls for astigmatism and trace rotation; these facilities are all described further in the next chapter. A feature of the DT12-5 is the wide-angle c.r.t. specially developed for the instrument by Thorn-Brimar. Its short overall length results in a compact instrument only 31 cm deep from front to back. The medium persistence mono-accelerator c.r.t. works at 2 kV final anode voltage giving a bright clear trace, and has a full 8 × 10 cm internal graticule for freedom from parallax.

The DT12-5 is only one of the Farnell range of oscilloscopes but it is fairly typical of a wide range of basic oscilloscopes available from many manufacturers. Some may have one or two facilities not found on the DT12-5 and vice versa, and like the DT12-5 most are (within the limits of this basic class of instruments) good value for money.

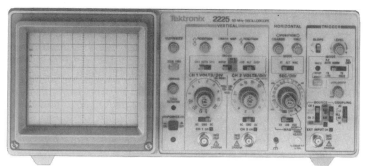

Figure 2.5 The Tektronix 2225 50 MHz two-channel oscilloscope features alternate magnification facility for performing waveform expansion and detailed signal analysis (courtesy RS Components Ltd)

13

3

Advanced real-time oscilloscopes

Entirely at the other end of the price-range from the basic type of oscilloscope described in Chapter 2 is the advanced oscilloscope. This typically has a host of features not found on a basic oscilloscope, and may be a mainframe plus plug-in system or a stand-alone scope. The latter type is often described as 'portable', to distinguish it from the former. The really advanced end of the oscilloscope market is shared between a number of manufacturers, possibly as many as a dozen, almost certainly less than a score. Nevertheless, however many facilities an oscilloscope manufacturer's top-of-the-range products may have, what really marks out the men from the boys is bandwidth. Few indeed are the manufacturers of real-time oscilloscopes with a Y bandwidth in excess of 200 MHz. Yet in high-speed computers 20 and 50 MHz clock rates or even higher are by no means uncommon, while in analogue systems a frequency response extending virtually up to the lower end of the u.h.f. band often reveals circuit problems, such as parasitic oscillations, that would otherwise pass unnoticed unless a spectrum analyser were to hand.

As in Chapter 2, an oscilloscope representative of its class is taken as an example, and its facilities discussed in detail. The oscilloscope chosen for this purpose is the Tektronix model 2465A *(Figure 3.1)*. This is a stand-alone instrument, requiring no plug-ins. Its facilities are comprehensive, and the following description covers nearly all of the points relevant to high-performance oscilloscopes. At the end of the chapter, however, reference is made to mainframe plus plug-in systems. These are more versatile than stand-alone oscilloscopes but generally work out more expensive for the same performance.

Stand-alone oscilloscope

The 2465A will first be described in its basic form, that is without any of the various options available. The 2465A operates from 115 V nominal (90 – 132 V) or 230 V nominal (180 – 250 V) a.c. mains (48 – 440 Hz) only, the maximum power consumption being 120 W. As

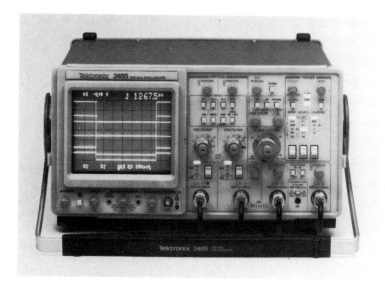

Figure 3.1 The original 300 MHz 2465 shown above has been replaced by the 2465A (see front cover) which now provides 350 MHz bandwidth at the probe tip. In addition, AUTO SET UP, SAVE and RECALL buttons speed measurements on new signals and previously encountered signals respectively. The AUTO button also allows one to step through up to 30 previously stored instrument set ups, held in non-volatile memory, for rapid execution of routine tests. The 2400 range also includes the 150 MHz 2445A, the 250 MHz 2455A and the 350 MHz 2467 with MCP c.r.t. (see Figure 7.11) (courtesy Tektronix UK Ltd)

one would expect, the 2465A has all the facilities found on the Farnell DT12-5 described in Chapter 2, though sometimes differently labelled. The facilities offered by the 2465A are so extensive that it is not possible in the confines of this chapter to describe them all in full detail; they greatly surpass the capabilities of the 475A described in the first edition of this book in 1981.

Power and display controls

On the left-hand side of the instrument, underneath the screen of the

15

c.r.t. with its 8 x 10 cm graticule, is a group of controls mainly concerned with the c.r.t. display. The leftmost of these is the 'intensity' knob, which controls the brightness of the trace(s), but not of the readout display of scale calibration factors. Next to this is 'beam find', a push button which, when held in, compresses the display to within the graticule area. This aids the operator in locating offscreen displays. The next control is the 'focus' knob and to the right of this is the 'trace rotation' control. This screwdriver-adjusted preset control can be used by the operator to align the c.r.t. trace with the horizontal graticule lines. Once adjusted it does not require readjustment during normal operation of the instrument. To the right of the trace rotation control and centrally placed beneath the screen is the 'readout intensity' control knob. Set to mid-travel, all readout messages are invisible. Rotating the knob to the left increases the intensity of any applicable readout messages, which include indications such as ↓ (Channel 2 inverted indicator), A1 (or whatever, indicating timebase A in use, triggered from Channel 1), BWL (bandwidth limit indicator) and 50 Ω overload indicator, etc. Rotating the knob to the right similarly increases the brightness of any readout messages and additionally displays vertical and horizontal scale factors, see *Figure 4.9*.

To the right of readout intensity is another screwdriver operated preset control. This is 'astig' (astigmatism) which is set so that the spot that writes the traces on the screen is circular, as described in Chapter 7. Next to astig is the 'scale illum' control knob which controls the brightness of the graticule illumination. The rightmost control of this group is the 'power' switch, a push button: press in for on, press again for off. An indicator in the button shows green when on, black when off. Front channel settings that were unchanged for at least 10 s prior to power-off will be remembered and reinstated when power is reapplied.

Like the oscilloscope described in the last chapter, the 2465A has an internal screen graticule for freedom from parallax errors. The graticule includes dotted horizontal lines at $2\frac{1}{2}$ divisions above and below the centre line, to facilitate rise and fall time measurements as illustrated in *Figure 8.4(c)*. In addition to the internal graticule, a blue tinted filter is fitted in front of the c.r.t. and a spare clear filter is supplied.

Vertical controls

The Y amplifier controls are located to the right of the c.r.t. screen. One of these, 'Ch1 or X' (Channel 1), doubles as the X input in XY mode. The Channel 1 input has a specially modified BNC input connector. This senses when the lead connected to it is one of the x 10 attentuator probes supplied with the instrument, automatically adjusting the deflection factor displayed on the c.r.t. readout to indicate the true deflection factor at the probe tip. Above the Channel 1 input socket is its input coupling switch and next to this a column of coupling mode indicator lights. From the top, these indicate 'a.c.,' 'gnd,' 'd.c.,' 'gnd' (all at 1 MΩ input resistance) and, at the bottom, 50 Ω. Pressing the upper button of the input coupling switch lights the next higher indicator and selects the corresponding input condition; similarly, pressing the lower button selects the next lower light and input condition. With a.c. selected, the lower frequency limit (−3 dB point) is 10 Hz or less. In the two '1 MΩ gnd' positions, the input to the Y amplifier is shorted (convenient when setting the trace 0 V reference level to mid-screen, for example) although the input resistance remains at 1 MΩ. The 50 Ω position is d.c. coupled and presents a 50 Ω resistive termination at the input socket. If a signal is applied which exceeds the permissible dissipation in the 50 Ω termination, the input coupling will automatically revert to 1 MΩ and a c.r.t. warning message readout will indicate the overloaded condition.

Above the input coupling switch is the 'volts/div' knob. This has 11 positions, providing, on the 2465, deflection factors from 2 mV/div in a 1 – 2 – 5 sequence anticlockwise, up to 5 V/div, all at the instrument's full 300 MHz bandwidth except for the 2 mV/div range, which is only specified to 100 MHz. Extended bandwidth is one of the most notable improvements incorporated in the 2465A. This instrument provides a bandwidth of 350 MHz at the probe tip on all ranges down to and including the 2 mV/div range. Concentric with volts/div is a small 'var' knob, which increases the deflection factor (anticlockwise) by up to 2.5:1 and causes a greater-than (>) sign to appear in front of the associated volts/div readout display on the screen. Above the volts/div switch, at the top of the panel, is a vertical shift control labelled '↕ position'. The Ch2 (Channel 2) input facilities are identical to those of Channel 1. If the Channel 1 and Channel 2 input signals are both a.c. coupled and if the upper buttons of both coupling switches are pushed together, the instrument automatically performs a d.c. balance of the Channel 1 and 2 vertical circuits. This is very convenient when using Channel 1 and 2 as a balanced floating input. Balanced measurements and CMRR are covered further in Chapters 4 and 5.

Two further Y inputs are provided, 'Ch3' and 'Ch4', via BNC input connectors. Like the Channel 1 and 2 inputs, the c.r.t. readout display of Channel 3 and 4 input deflection factors is automatically adjusted to read the true deflection factors when a × 10 probe is connected. Associated with the Channel 3 input is a vertical shift position control and a locking push button providing a deflection factor of 0.1 V/div (out) or 0.5 V/div (in). Input coupling on Channel 3 is d.c. only. Channel 4 facilities are identical to those of Channel 3. Channels 3 and 4 are most useful as digital signal and trigger signal input channels, given their limited choice of deflection factors.

Between the Channel 3 and 4 inputs is a calibrator connector which provides a 0.4 V p–p (peak-to-peak) squarewave into a 1 MΩ load, or 0.2 V p–p into 50 Ω or 8 mA p–p into a short circuit. It is useful for checking voltage and current probes but also serves to check vertical deflection accuracies and sweep speeds and delays. To this end, the repetition rate of the square wave changes with the setting of the A sec/div switch.

Control of the various Y inputs is exercised by a group of 8 latching push buttons situated between the volts/div switches and the vertical position controls of Channels 1 and 2. Depressing any combination of the four buttons Ch1 to Ch4 causes the appropriate traces to be displayed. Pressing the 'add' button brings up an additional trace displaying the sum of the inputs to Channels 1 and 2. Pressing 'invert' causes the Channel 2 trace to be inverted, i.e. positive voltages produce a downward deflection and vice versa. In this case, the add trace displays the difference of the Channel 1 and 2 signals, thus providing in effect a balanced floating input. Any combination of the five signals can be displayed by pressing in the appropriate push buttons, but if none of them is latched in, then the Channel 1 signal will be displayed. When multiple channels are selected, they are displayed sequentially in the following 'pecking order': Ch1, Ch2, add, Ch3, Ch4. The '20 MHz bw limit' switch reduces the upper −3 dB bandwidth of the vertical deflection system when latched in. The eighth push button, 'chop out:alt', controls the instrument's mode of displaying multiple traces. Chop (latched in) causes the sequential display of short segments of each trace in priority order. The chopping frequency is between 1 and 2.5 MHz, desynchronized from the sweep repetition frequency, to minimize waveform breaks when viewing repetitive signals. Thus the chop mode may be used even when viewing high-frequency signals, although it is most appropriate at lower timebase settings. Out:alt (released) causes multiple traces to be displayed sequentially in priority order. It is thus more appropriate at moderate timebase speeds upwards, since at low speeds one would see up to five (or even ten) traces being written one after the other, whereas they would appear simultaneous in chop mode. If both A and B sweeps are being displayed (see trace sep control, below) then the A and B sweep for one trace will be displayed in that order before switching to the next priority trace.

Horizontal and Δ measurements

Compared to previous generation oscilloscopes such as the 475A, the facilities offered by the 2465A model are so extensive that the following description is necessarily somewhat abridged.

The timebase speed is controlled by the concentric 'A and B sec/div' knobs. The A sec/div switch selects any of 25 calibrated A sweep speeds from 0.5 s per division to 5 ns per division, or delay ranges from 5 s to 100 ns, in a 1 – 2 – 5 sequence. Extreme counter-clockwise rotation selects the X–Y display mode. In X–Y, the signal applied to the Ch1 or X input connector drives the horizontal deflection system. The B sec/div switch selects any of 22 calibrated B sweep speeds from 50 ms per division to 5 ns per division in a 1 – 2 – 5 sequence. This switch also controls the horizontal display mode switching as follows. When the A and B sec/div switches are set to the same sweep speed and the B sec/div knob is pushed in, the two knobs are locked together and rotate as one; in this position, only the A sweep is displayed on the c.r.t. The 'pull-inten' mode applies when the A and B sec/div switches are set to the same sweep speed, with the B sec/div switch pulled out. In this mode, the B sweep is not displayed but runs much faster than the A sweep. The A sweep is intensified for the duration of the B sweep time. The position of the intensified zone on the A sweep indicates the delay time between the start of the A and B sweeps. Its position is controlled by the 'Δ ref or dly pos' control. For single trace displays, when either the delta time (Δt) or the reciprocal delta time (1/Δt) function is activated, a second intensified zone will appear on the A sweep (if the B trigger mode is set to 'run aft dly' or to 'trig aft dly' with a suitable trigger signal applied). The position of the first (reference) zone is controlled by the 'Δ ref or dly pos' control as before, and the position of the delta zone by the Δ control. With multiple traces, the operation is somewhat different.

In 'turn alt' mode, pulling the B sec/div knob to the out position, then turning it to a faster sweep speed (interlocking prevents it ever being set to a slower speed), produces the alternate (alt) horizontal display mode. The A sweep with an intensified zone will be alternately displayed with the B sweep, provided the B trigger mode is set either to 'run aft dly' or to 'trig aft dly' with a proper B triggering signal occurring before the end of the A sweep. The alternate A and B traces will be superimposed unless separated with the trace sep control (see below). The position of the intensified zone on the A sweep indicates the approximate delay of the B sweep, and the length of the intensified zone the approximate B sweep duration set by the B sec/div switch. If either Δt or 1/Δt is also activated, intensified zones and associated B sweeps will be established in the same manner as described in 'pull inten'. Push B – pushing in the B sec/div switch when set to a faster sweep speed than A sec/div – presents only the B trace(s) on the c.r.t. display.

The 'sec/div var' control continuously varies the sweep speed between settings of either the A or the B sec/div switch. This control affects the A sweep speed when the A and B sec/div switches are locked together. When any of the delayed sweep horizontal modes are displayed, the control affects only the B sweep speed. Fully counter-clockwise rotation extends the sweep of the slowest A sec/div switch setting from

0.5 to 1.5 s per division. The c.r.t. readout displays the actual time per division scale factor for all settings of the var control. This control produces fine resolution over a portion of its range, after which it changes to coarse resolution. It re-enters the fine resolution range upon reversing the direction of rotation.

The trace sep control provides for vertical positioning of the B trace downward from the A trace when turn-alt horizontal display mode is selected. Starting from zero separation, counter-clockwise rotation moves the B trace downwards. When the push-B horizontal mode is selected with Δt or 1/Δt, the trace sep control provides for vertical positioning of the trace(s) associated with the Δ control.

The horizontal position control sets the horizontal position of the sweep display on the c.r.t. This control produces fine resolution over a portion of its range, like the 'sec/div var' control. The x 10 mag switch horizontally magnifies the portion of the sweep display positioned at the centre vertical graticule line by a factor of 10 when pressed in. When displaying A and B traces alternately (turn-alt mode), only the B trace is magnified. x 10 mag extends the fastest speed to 500 ps per division.

Voltage measurements on any waveform or portion thereof can be estimated from the vertical graticule divisions. However, much more accurate measurements are possible with the 2465A's voltage cursors. The ΔV switch activates the delta volts (ΔV) measurement function, when momentarily pressed in alone, and cancels any other delta measurement function in effect. In the A sweep mode, two horizontal cursors are superimposed on the display. The c.r.t. readout displays the equivalent voltage represented by the separation between the cursors. The position of one cursor is set by the 'Δref or dly pos' control and that of the other by the Δ control. With multiple channel displays, the deflection factor of the first channel selected in the display sequence determines the scale factor of the delta volts readout on the c.r.t. The delta volts readout is displayed as a percentage ratio in the following cases: (1) the channel determining the scale factor is uncalibrated, or (2) 'add' is displayed alone when the deflection factors of Channels 1 and 2 are not the same.

Pulse widths, delays etc. can be estimated from the horizontal graticule divisions, knowing the sweep speed. However, greater accuracy is provided by the 2465A's time cursors. The Δt switch activates the delta time function and cancels any other delta function when pressed: a second press cancels the delta time function. With the A trace, only, Δt superimposes two vertical cursors on the c.r.t. display. In any of the delay time horizontal modes, two separate delay times are established by the delta time function. One cursor position (or delay time) is set by the Δ ref or dly pos control, and the other is set by the Δ control. The c.r.t. readout displays either the time difference between the two delays or the equivalent time difference between the two cursors. If the 'sec/div var' control is not in the detent position (fully clockwise), Δt cursor difference on A trace only displays is expressed as a ratio, with five divisions corresponding to 100 per cent. For the delay-time horizontal display modes, the sec/div var control varies the B sweep scale factor as it is rotated, but it has no effect on the delay time readout.

Pressing the ΔV and Δt push buttons simultaneously activates the $1/\Delta t$ measurement function. The Δ ref or dly pos and Δ controls work as previously, but now the c.r.t. readout shows the reciprocal of the time difference measurement, with units being frequency (Hz, kHz, MHz or GHz). For A trace only displays, with the sec/div var control out of detent, the time difference between $1/\Delta t$ cursors is displayed in degrees of phase, with five divisions equal to 360 degrees. The greatest accuracy in time measurements is obtained when using the delayed sweep mode. The Δ ref or dly pos and Δ controls produce fine resolution over a portion of the range, like the sec/div var control.

The 'tracking-out:indep' switch selects either the tracking or independent mode for the Δ ref or dly pos control. When in the tracking mode (push button latched in), the difference between alternate delay times or cursors (in either time or volts measurement modes) does not change with rotation of the Δ ref or dly pos control, i.e. both delays or cursors move in sympathy (unless the limit of either is reached). If 'out:indep' is selected, each control adjusts only its own delay or cursor. In either case, the Δ cursor is always independently movable by the Δ control.

Triggering

The trigger 'mode,' 'source' and 'coupling' are each controlled by a pair of push button switches, each with its adjacent column of indicator lights. The switches control the lights and their associated functions in the same way as the Y input coupling switches, described earlier. The following description of the mode switch details the functions from the top indicator light downwards, and similarly for the source and coupling switches.

The mode switch selects the trigger mode of either the A or B sweep. The modes provided are as follows:

A Trigger Modes. 'Auto lvl' automatically establishes the trigger level on a trigger signal if present, or free runs the sweep. The level control covers a range between the positive and negative peaks of repetitive triggering signals.
Auto sweep free runs in the absence of a triggering signal. The triggering level changes only when the level control is adjusted to a new position. Norm (normal) - the sweep only runs when the selected triggering

signal crosses the level set by the level control (except under certain circumstances where this would be inconvenient).

Sgl seq (single sequence), when armed by pushing the mode switch lower button momentarily, causes the sweep to run one or more times to produce a single sweep of each of the traces selected for display. Each sweep requires a distinct A sweep triggering event. The ready indicator remains illuminated until the final trace in the sequence is completed. At the end of the sequence the c.r.t. readout is written once to present scale factors and other readout data, and scale illumination flashes on momentarily for oscilloscope photography purposes.

B Trigger Modes. 'Run aft dly' causes the B sweep to run immediately after the set delay time has elapsed.

'Trig aft dly' causes the B sweep to run on the first trigger event after completion of the set delay, provided the A sweep has not terminated in the meantime. Since the actual delay time is greater than the delay time setting, the c.r.t. delay time readout shows a question mark in this mode.

The source switch selects the trigger signal source for either the A or the B sweep. Indicators show the selection made, but do not illuminate for B triggering signals when 'run aft dly' is selected. The available trigger sources are as follows:

Vert (vertical) triggers the sweep on the displayed (single) channel. If more than one channel is selected for display, then the triggering source is determined by the setting of certain other controls.

Ch1, Ch2, Ch3, or Ch4 select the indicated channel as the triggering source.

Line (A trigger only) triggers the instrument in synchronism with the a.c. power source.

The coupling switch selects the method of coupling the triggering signal to the trigger generator circuits. The different types of coupling are as follows:

D.c. where all frequency components of the trigger signal are fed to the trigger generator.

Noise reject is similar to d.c. coupling but with larger hysteresis; it ignores low level noise riding on a waveform.

Hf rej is similar to d.c. but the response rolls off above 50 kHz.

Lf ref is an a.c. coupled position with the response below 50 kHz rolling off.

A.c. provides a roll off of frequencies below 60 Hz.

The A/B trig switch directs the mode, source, coupling, slope and level controls to either the A trigger or the B trigger, under the allowed switching conditions. Controls are normally directed to the A trigger when the A and B sec/div knobs are locked together (A sweep only) and to the B trigger otherwise (in trig aft dly mode). Pressing and holding in the A/B trig switch will direct the trigger controls away from their normal direction, but releasing the A/B trig switch will redirect the trigger controls back to the original triggers. The A/B trig switch also has certain other functions.

The level control sets the point on the waveform selected for triggering at which the timebase triggers. This control produces fine resolution for a portion of its range, like the sec/div control. When the A trigger mode is set to auto lvl, the effect of the level control is spread over the A sweep triggering-signal range from peak to peak. In this case, rotating the control to either extreme causes the triggering level to be redefined by the auto lvl circuitry.

The slope switch selects the slope of the signal that triggers either the A sweep or the B sweep. + and − indicators show the slope selected (except in the case of B sweep with run aft dly selection). The 'A swp trig'd' indicator illuminates to indicate that the A sweep has been triggered but goes out again if further triggers are not forthcoming. The ready indicator illuminates when sgl seq mode is selected and the A sweep is armed and awaiting a trigger. It goes out when all the traces selected for display have been completed.

The hold-off control varies the amount of 'hold-off' time between the end of a sweep and the earliest time it can be retriggered. The ability to obtain stable triggering on some aperiodic signals is improved using this control, as explained with the aid of an example in Chapter 5. In the 'B ends A' position (fully clockwise), trigger hold-off time is reduced to minimum and the A sweep terminates immediately at the end of the B sweep. This enables the fastest possible sweep repetition rate at slow A sweep speeds.

Back panel

It is usual on high performance portable oscilloscopes to banish certain less frequently used facilities such as the mains fuse to the rear, saving valuable space on the instrument's front panel: the 2465A follows this time-honoured tradition. The two-position mains selector switch provides the voltage operating ranges aleady mentioned, and is situated beneath the receptacle for the detachable power cord. The power cord is normally secured to the rear panel by a cord-set-securing clamp. The other end of the three-wire power cord is fitted with the type of three-pin plug required by the user. Option A2(UK) specifies a power cord fitted with a British Standard 'ring main' plug to BS1363, fitted with a 6 A fuse, but other options cover virtually all types of mains connector.

At the bottom of the rear panel are four BNC sockets with the following functions. The 'Ch2 signal out' connector provides a replica of the Channel 2 signal at a level of 10 mV per division into 50 Ω or 20 mV per division into 1 MΩ. By connecting this signal back into the

Channel 1 input, the Channel 2 input sensitivity can be increased by a factor of × 2 or × 4 (with bandwidth restrictions). The 'ext Z-axis in' connector provides a means of modulating the trace intensity, positive going voltages reducing the intensity. A typical application of this facility is described in Chapter 5. The two remaining BNC connectors provide the A and B gate out signals. These positive going TTL compatible signals of length equal to the A and B sweeps can be very useful in a test set-up for controlling some function of the circuit under test in sympathy with the display, or even for initiating the function if the sweep is allowed to free run.

Options

The power cord option has already been mentioned, but apart from this several other options are available on the basic model 2465A. Option 05 (also available on the 150 MHz model 2445A) specifies television waveform assessment capabilities, including facilities for triggering from field 1, field 2 or alternate fields of an interlaced format, and from any specified line number. Other features include a back porch blanking level clamp and CCIR and NTSC graticules. Option 10 (also available on the 2445A) specifies a GPIB (general purpose interface bus) implementation, covering 11 of the standard interface functions.

Option 22 specifies two more P6131 passive probes in addition to the two supplied as standard with the oscilloscope, while Option 11 provides two power connectors on the rear panel, furnishing the supplies for Tektronix active oscilloscope probes. Active probes are described in Chapter 4. Option 1R specifies the rack mounting version.

In addition to the basic model 2465A, with or without options, there are three enhanced models, namely, 2465A CT, 2465A DM and 2465A DV (see cover picture). All include the GPIB, a counter/timer/ trigger word recognizer which provides crystal timebase measurement accuracy and word, event and Boolean triggers, and come with four probes as standard. Additionally, the DM and DV models have built in, on top of the basic instrument, a fully programmable $4\frac{1}{2}$ digit autoranging DMM (digital multimeter) with d.c., a.c. (rms) volts and current ranges, dBm, dBV, resistance and temperature measurement facilities. All measurements can be averaged and smoothed for c.r.t. display or transmission over the GPIB. The top-of-the-line DV model also includes the television waveform facilities described under option 05.

Mainframe plus plug-in oscilloscopes

Mainframe and plug-in systems are designed for bench operation from a.c. mains supplies rather than portable use. A good example is the long-established Tektronix 7000 series, see *Figures 6.7, 6.13* and *6.19*. *Figure 3.2* shows the latest generation of Tektronix mainframe oscilloscopes.

The advantage of the plug-in plus mainframe format is economy, since if a different facility is needed it can be had for the cost of a plug-in, whereas otherwise a complete new oscilloscope would be required. On the other hand, only one person at a time can use the mainframe, so usually at any one time capital is tied up in unused plug-ins. Most large electronics laboratories therefore seek to strike a balance, with some mainframe oscilloscopes plus a wide range of plug-ins for versatility, and some stand-alone oscilloscopes (often described as portables) for economy.

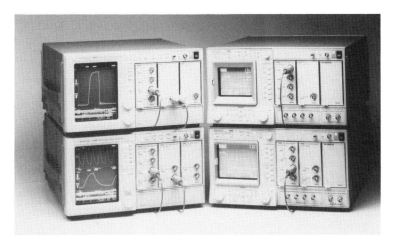

Figure 3.2   The new Tektronix 11000 series mainframe plug-in oscilloscopes. Both of the 11400 series digitizing oscilloscopes (left-hand side) have a 20 M samples per second maximum sampling rate, with a repetitive (equivalent time) bandwidth of 500 MHz for the 11401 and 1000 MHz for the 11402 (upper). The 1130 series analogue scopes (right) offer a maximum bandwidth of 400 MHz (11301, lower) and 500 MHz (11302), depending on the vertical plug-in selected. The 11302 uses the Tektronix microchannel plate c.r.t. (see Figure 7.11), providing a visual writing speed of 10,000 divisions/μs. Note that unlike the earlier 7000 series mainframes, 11000 series scopes require no timebase plug-in, having full timebase facilities built-in (courtesy Tektronix UK Ltd)

4

Accessories

We have examined a variety of oscilloscopes in the preceding chapters: the simple, the advanced and the modular. All are capable of examining waveforms just as they stand; simply connect the circuit whose waveform you wish to examine to the Y inputs and the waveform will appear on the screen (assuming the controls are suitably set).

Actually, it is not quite that simple. Although the Y input of an oscillocope has a very high impedance, in many cases its effect upon the circuit to which it is connected is not entirely negligible. The standard Y input resistance is 1 MΩ, and the input capacitance is generally within the range 15–40 pF depending upon the particular make and model. With such a high input impedance, hum pick-up on the input lead would often be a problem when examining small signals in high-impedance circuits unless a screened lead were used. However, one metre of screened lead could easily add another 50–100 pF to the oscilloscope's input capacitance; on the other hand, trying to connect the circuit under test directly to the input connector of the oscilloscope with negligible lead lengths is always tedious and often impossible. The usual solution to this problem is a passive divider probe, and this is the first accessory at which we shall look.

Passive divider probes

Experience shows that, to connect an oscilloscope to a circuit under test, a lead about one metre in length is usually convenient, screened to avoid hum pick-up when working on high-impedance circuits.

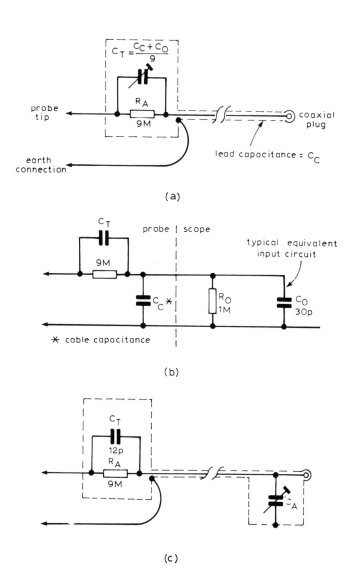

Figure 4.1 (a) Circuit diagram of traditional 10:1 divider probe. (b) Equivalent circuit of probe connected to oscilloscope. (c) Modified probe circuit with trimmer capacitor at scope end (courtesy *Practical Wireless*)

Even a low-capacitance cable has a capacitance of about 60 pF per metre, so a metre of cable plus the input capacitance of the oscilloscope would result in about 100 pF of input capacitance. The purpose of a 10:1 passive divider probe is to reduce this effective input capacitance to around 10 pF. This is a useful reduction, bearing in mind that at even a modest frequency like 10 MHz, the reactance of 100 pF is as low as 160 ohms.

Figure 4.1(a) and *(b)* show the circuit diagram of the traditional type of oscilloscope probe, where C_O represents the oscilloscope's input capacitance, its input resistance being the standard value of 1 MΩ. The capacitance of the screened lead plus the input capacitance of the oscilloscope form one section of a capacitive potential divider. The trimmer C_T forms the other, and it can be set so that the attenuation of this capacitive divider is 10:1 in volts, which is the same attenuation as provided by R_A (9 MΩ) and the 1 MΩ input resistance of the oscilloscope. When this condition is fulfilled, the attenuation is indepedent of frequency—*Figure 4.2(a)*. Defining the cable plus oscilloscope input capacitance as C_E, i.e. $C_E = (C_C + C_O)$, C_T should have a reactance of nine times that of C_E, i.e. $C_T = C_E/9$. If C_T is too small, high-frequency components (e.g. the edges of a square wave) will be attenuated by more than 10:1, whereas the attenuation of the steady level will still be 10:1, resulting in the waveform of *Figure 4.2(b)*. If C_T is too large, the result is as in *Figure 4.2(c)*.

The input capacitance of the oscilloscope C_O is invariably arranged to be constant for all settings of the Y input attenuator, so that C_T can be adjusted by applying a square wave to the oscilloscope via the probe using any convenient Y sensitivity. Many oscilloscopes provide a square wave output on the front panel specifically for setting up passive divider probes. Such probes most commonly provide a division ratio of 10:1, but other values (e.g. 100:1) are sometimes found and some 10:1 probes have provision for shorting R_A and C_T to provide a 1:1 ratio. When using such a probe in the 1:1 mode, the capacitive loading of the circuit under test is of course ten times as great as in the 10:1 mode, and its use is therefore confined mainly to lower frequencies.

The circuit of *Figure 4.1(a)* provides the lowest capacitive circuit loading for a 10:1 divider probe, but has the disadvantage that 90 per cent of the input voltage (which could be very large) appears across the variable capacitor C_T. Some probes therefore use the circuit of *Figure 4.1(c)*; C_T is now a fixed capacitor and a variable shunt capacitor C_A is fitted, which can be set to a higher or lower capacitance to compensate for oscilloscopes with a lower or higher input capacity respectively. Now only 10 per cent of the input voltage appears across the trimmer, which can also be conveniently located at the oscilloscope end of the probe lead, permitting a smaller, neater design of probe head.

Even if a 10:1 passive divider probe is incorrectly set up, the rounding or pip on the edges of a very low-frequency square wave, e.g. 50 Hz, will not be very obvious, because with the necessary slow timebase speed the square wave will appear to settle very rapidly to the positive and negative levels. Conversely, with a high-frequency square wave, say

1 MHz, the division ratio will be set solely by the ratio C_E/C_T. Waveforms as in *Figure 4.2* will be seen at frequencies of one or two kilohertz.

At very high frequencies, where the length of the probe lead is an appreciable fraction of a wavelength, reflections occur, since the cable is not terminated in its characteristic impedance. For this reason, oscilloscope probes often incorporate a resistor of a few tens of ohms in series with the inner of the cable at one or both ends, or use a special cable with an inner made of resistance wire. Such measures are necessary in probes that are to be used with oscilloscopes of 100 MHz bandwidth or more.

Special X10 passive divider probes are available for use in pairs with an oscilloscope with a Y1–Y2 facility (Channel 1 + Channel 2, with Channel 2 inverted.) By effectively making both R_A and C_T adjustable (*see Figure 4.1*) the gain of the oscilloscope's two Y channels can be equalized at both low and high frequencies. For example, two Tektronix P6055 probes can provide 20,000:1 CMRR (common mode rejection ratio) from d.c. to 1 KHz, derating to 100:1 at 20 MHz.

Not only the use of passive divider probes but also their theory has been covered in this chapter (rather than in Chapter 8) because they are by far the commonest—and to that extent the most important—oscilloscope accessory. Many a technician (and chartered engineer too) has wasted time wondering why the amplitude of a 1 MHz clock waveform, for example, was out of specification, only to realise eventually that the 10:1 passive divider probe he was using was not correctly set up for use with that particular oscilloscope!

Active probes

The reduced capacitive circuit loading provided by the passive divider probe is dearly bought, the price being a reduction in the sensitivity of the oscilloscope, usually by a factor of 10. An active probe can provide a 1:1 ratio, or even in some cases voltage gain, while still presenting a very low capacitive load to the circuit under test.

This is achieved by mounting a small unity-gain buffer amplifier with a high input impedance and a low output impedance actually in the probe head. In some cases the probe head has two leads, a coaxial cable to the Y input socket and a power supply lead to an accessory power socket on the oscilloscope or to a special separate free-standing probe power-supply unit. With the simple arrangement described, the maximum signal that can be usefully applied to the probe is obviously limited by the input voltage swing that the probe head amplifier can handle. This can usually be increased by the use of 10:1 or 100:1 divider caps, clipped on to the active probe's input.

The extensive Tektronic range of active probes includes types with bandwidths up to 1.5 GHz, with a rise time of 230 ps.

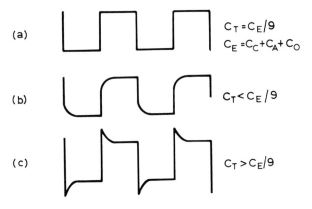

(a) $C_T = C_E/9$
$C_E = C_C + C_A + C_O$

(b) $C_T < C_E/9$

(c) $C_T > C_E/9$

Figure 4.2 Displayed waveforms with probe set up (a) correctly, (b) under-compensated, (c) overcompensated

Current probes

The probes described so far, both active and passive, are designed for the measurement of voltage waveforms. However, probes are also available that measure current waveforms. There are passive current probes, but these usually have low sensitivity and a limited frequency response that does not extend down to d.c., though they can be useful where these limitations are not important.

Current probes usually have a slotted head, the slot being closed by a sliding member, after slipping in the wire carrying the current to be measured. There is thus no need to break the circuit to thread the wire through the probe. Current probes produce an output voltage waveform identical to the waveform of the current flowing in the wire passing through the slot in the probe head.

Current probes with a frequency response down to d.c. are usually active types, though most of the electronics is contained in an interface box to which the probe connects and which has an output lead to the Y input socket of the oscilloscope. Passive a.c. only probes can be used plugged directly into an oscilloscope, though in this mode the low-frequency cut-off point, depending on the particular probe, may be anywhere in the range from under 100 Hz to 1 kHz or more. However, special amplifiers are available to interface the probe to an oscilloscope; these not only increase the sensitivity of the probe, but extend its low frequency cut-off point downwards by a factor of about ten.

30

Viewing hoods

Modern oscilloscopes are generally entirely satisfactory when displaying a repetitive waveform, say a sine wave or a pulse train with a mark/space ratio near unity. However, if the pulse is narrow and the repetition frequency low, for example a 1 μs wide pulse occurring once every 1 ms, many oscilloscopes on the market (especially the less expensive ones) will not produce a bright enough picture of the pulse to be useful. For if the timebase speed is set to 1 μs per division, then with the usual ten horizontal divisions the spot is blanked for 99 per cent of the time and only drawing a trace for the remaining 1 per cent. The trace is therefore so dim as to be invisible, owing to the reflection of ambient room lighting from the tube face and graticule.

A really good viewing hood enables the user to view the screen while shutting out all ambient light. When the eyes become dark-adapted, even a very faint trace can be seen. The author recalls that with the aid of its snug-fitting viewing hood the Tektronix 545 oscilloscope, designed in the 1950s, was capable of displaying a 1 μs pulse occurring once a second. Nowadays, of course, one would probably wheel up a storage oscilloscope, but a viewing hood is after all much cheaper!

Oscilloscope cameras

Besides viewing a waveform, one may wish to make a permanent record of it. Although tracing paper and a steady hand may suffice for simple waveforms, an oscilloscope camera is the usual answer. Tektronix and Hewlett Packard both produce cameras to fit their respective oscilloscopes, but many of the smaller oscilloscope manufacturers do not.

Probably the best-known manufacturer (at least in the UK) of cameras to fit virtually any make of oscilloscope is Shackman Instruments Ltd. *Figure 4.3* shows Shackman camera types Super 7 and 7000, the latter complete with an adaptor that enables it to fit on to an Advance oscilloscope type OS1000. It can be seen that the Super 7 camera also includes a viewing hood so that the trace on the screen can be viewed immediately prior to taking the picture. A shutter enables this oblique viewing facility to be blacked out, excluding all extraneous light.

Oscilloscope cameras can use film of a sensitivity much greater than that of the human eye, so that a very narrow pulse occurring as a single event can be photographed even though invisible to the eye. *Figure 4.4* shows the density of the image on a developed film as a function of quantity of light falling on it during the exposure. The exposure axis is arbitrary, unity corresponding to that illumination which will produce a mid grey. Gross overexposure can only result in the maximum density

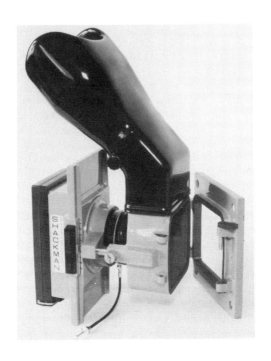

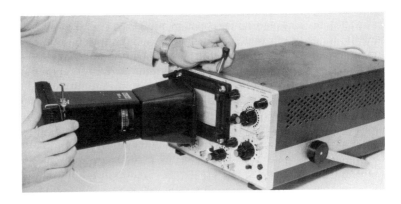

Figure 4.3 Oscilloscope cameras from the Shackman range. (a) Super 7 camera with $f/1.9$ lens and adaptor; a range of over sixty adaptors is available, matching the camera to over 350 different oscilloscopes. (b) Inexpensive model 7000 with $f/3.5$ lens, shown fitted to a Gould Advance oscilloscope type OS1000 (courtesy Shackman Instruments Ltd)

image; gross underexposure results in no image at all. The curve of image density against exposure is in fact an S-shaped curve familiar to all photographers.

If the trace we wish to record is so faint that even the fastest film available will be underexposed, it may still be possible to photograph it. This is achieved by deliberately applying an exposure evenly across the whole negative before, during or after photographing the single-event trace. The technique, known as 'pre-', 'simultaneous-' or 'post-fogging', moves the effective exposure from the flat bottom (underexposed) end of the S curve up to the steepest central part. It results in the trace appearing on the negative as a slightly darker line against a mid-grey background (on the positive print, whether produced separately later or 'instantly' as with the Polaroid system, the trace will appear as a lighter grey line on a darker grey background). The greatest improvement is obtained with simultaneous fogging. Photography using the fogging technique is a means of increasing the effective 'writing speed' of an oscilloscope. (Writing speed is covered more fully in Chapter 7.) 'Writing-speed enhancers' using the fogging technique are available for use with various oscilloscope cameras, e.g. the Tektronix model in *Figure 4.5*.

The cameras described so far capture the picture on the screen of the c.r.t. and record it on film. In place of film, the Tektronix DCS (Digitizing Camera System) uses a CCD video sensor. Charge coupled devices can be used for a number of purposes, another application being described in Chapter 6. The DCS uses a CCD array video camera (see *Figure 5.5*), with one part in 512 resolution in both the vertical and horizontal axes. Each pixel (display point) is digitized on a 256 level grey scale and the complete video data relating to one exposure transferred to mass storage in an IBM PC, XT, AT or compatible personal computer, in about a third of a second. In 600 ms it can be formatted and displayed on the computer's monitor screen, for examination and further processing – zoom, differentiation, smoothing etc.

On oscilloscopes with a microchannel plate c.r.t. (see Chapter 7), the DCS captures both repetitive and transient signals at the full bandwidth of the scope. On other Tektronix analogue oscilloscopes it is usable to 60% or more of the scope's bandwidth, depending on writing speed considerations. The DCS is also compatible with various models of other makes of oscilloscope, including B&K, IWATSU and KIKUSUI.

Oscilloscope calibrators

These come in varying degrees of complexity. One of the simpler types is often actually incorporated within the better class of oscilloscope. This was particularly necessary in earlier years before the advent of the

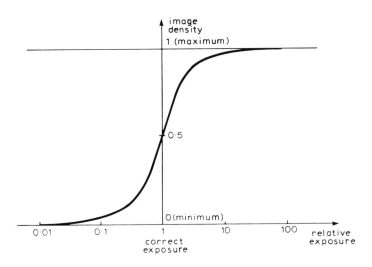

Figure 4.4   Photographic S curve of image density versus exposure

Figure 4.5   Tektronix C51 oscilloscope camera fitted with writing-speed enhancer (not visible in this view) (courtesy Tektronix UK Ltd)

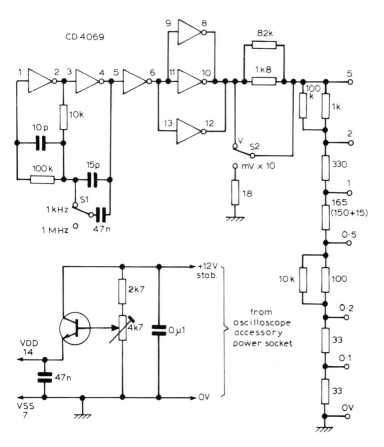

Figure 4.6 Simple oscilloscope calibrator. The 4.7 kΩ preset is adjusted to give a 5 V peak-to-peak output; 1 kHz and 1 MHz frequencies are nominal (courtesy *Practical Wireless*)

transistorised oscilloscope, as valves are subject to a steady decline in their characteristics the longer they are used.

Owing to their usefulness, calibrators of varying degrees of complexity are still incorporated in more expensive oscilloscopes, often with a choice of several accurate voltage levels at a frequency that is nominally 1 KHz to within a few per cent. Also, in some cases a metal loop projecting from the front panel carries a square wave of current of an accurate value to enable current probes to be calibrated.

As a separate accessory, a typical oscilloscope calibrator provides a clean square-wave output with an accurate peak-to-peak voltage swing or, more usually, a choice of peak-to-peak swings. There may also be a choice of frequencies; if only one is provided it is usually of the order of 1 KHz to enable passive divider probes to be set up, as described earlier. The choice of output voltage swing enables the Y deflection factor of an oscilloscope to be checked on each range, or at least on all the more sensitive ranges.

The more expensive calibrator, such as might be found in an instrument calibration laboratory, also offers a wide range of accurate standard square-wave frequencies, say 10 Hz to 10 MHz with intermediate steps in a 1:2:5 sequence. This enables the accuracy of an oscilloscope's timebase ranges to be checked.

Figure 4.6 is the circuit diagram of a simple oscilloscope calibrator that the author designed for use with the *Practical Wireless* Purbeck oscilloscope. It is intended to be powered from the 12 V d.c. stabilised supply available at the Purbeck's front panel accessory power socket, but could be used from any suitable power source, even a PP9 9 V battery. The preset potentiometer is set so that the maximum output swing is exactly 5 V peak-to-peak.

Special graticules etc.

All oscilloscopes nowadays incorporate a graticule, ruled in square divisions, usually ten horizontal divisions by six (or more commonly eight) vertical ones. On oscilloscopes with a cathode-ray tube of 13 cm (5 inches) diameter (or diagonal in the case of the increasingly popular rectangular-screen tubes), these divisions are generally centimetre squares. In addition to the square graticule rulings, many oscilloscopes have horizontal dotted lines across the graticule, $2\frac{1}{2}$ divisions above and below the centre line. If the top and bottom edges of a square wave are aligned with these, the rulings 2 divisions above and below the centre line intersect the edges of the square wave at the 10 per cent and 90 per cent points, making it easy to measure the rise and fall times.

More expensive oscilloscopes have variable-intensity edge lighting, which helps to make the graticule divisions stand out. In addition, there is generally a transparent sheet of perspex or similar material in front of the tube, tinted the same colour as the c.r.t.'s trace. This tinted sheet may have graticule divisions marked upon its rear surface. (Alternatively, in some oscilloscopes, the graticule is printed on the inside of the c.r.t. screen before the phosphor is applied; this completely eliminates parallax between the trace and the graticule, but of course makes it impossible to change the latter.)

The purpose of the tinted sheet is to improve the contrast between the trace and the rest of the screen. The brightness of the trace is reduced somewhat by the tinted sheet, but the trace's light only has to pass through the sheet once. Ambient light, on the other hand, is

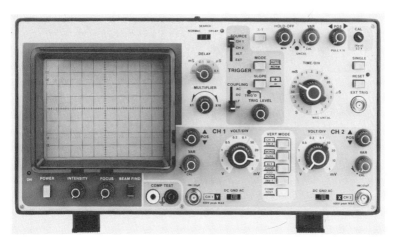

Figure 4.7 The Beckman Industrial 9020 20 MHz two-channel oscilloscope features sweep delay and a component tester (courtesy RS Components Ltd)

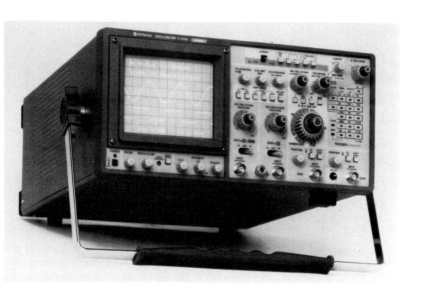

Figure 4.8 The Hitachi range of oscilloscopes extends from single channel 15 MHz types through 40Msample/sec digital storage models to the four channel 100 MHz V-1100A above (courtesy Hitachi Denshi (UK) Ltd)

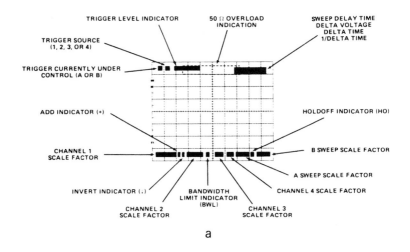

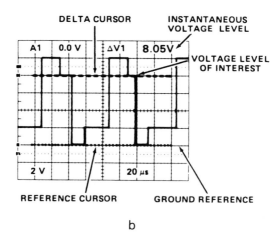

Figure 4.9 (a) The scale-factor and warning readouts provided on the c.r.t. display of the 2465A oscilloscope. (b) A typical measurement application, with the Delta Volts function in use (courtesy Tektronix UK Ltd)

attenuated as it passes through the tinted sheet before being reflected by the c.r.t. screen, and attenuated again as it passes out, resulting in improved contrast. The improvement is even greater for colours different from that of the trace.

Increasingly, a neutral grey-tinted sheet is used in place of one the same colour as the trace. The material is a special plastic sheet with the property of circularly polarising light that passes through it. The trace

38

is little attenuated, because it is initially unpolarised and only passes through the filter once, but ambient light falling on the screen of the c.r.t. is circularly polarised. After it has been reflected, its circular polarisation is largely of the wrong hand to pass back through the filter, resulting in much improved contrast.

In particularly bright surroundings a mesh filter can be used. This is a fine metal mesh finished in matt black; it reduces the brightness of the trace by about a quarter, but provides high attentuation of ambient-light reflections.

Special graticules have been produced for fitting to an oscilloscope in place of the standard one. A typical example is a graticule with nominal and limit markings for a sine-squared pulse-and-bar test wave-form, used for testing television equipment for response time and differential gain and phase. Smith chart and polar graticules are also available, but these are generally used with a special-purpose oscillo-scope display forming part of a network analyser.

Mains isolation

The Y input sockets on an oscilloscope normally have their outer screens connected to the instrument's metalwork and to the earth wire in the mains lead. Thus the input as it stands cannot be connected to circuitry which is at an elevated potential with respect to mains earth, for example live-side components in a direct off line switchmode power supply. Hence the highly deprecated and very dangerous practice of disconnecting an oscilloscope's earth lead. However, under specified conditions, safety standards do permit indirect grounding as an alterna-tive to direct grounding. All of the grounding requirements apply, except that the grounding circuit need not be completed until the available voltage or current exceeds a prescribed amount. The Tektronix A6901 Ground Isolation Monitor fits between an oscilloscope and the mains, and continuously monitors the voltage on the instrument's case/metalwork. The latter is permitted to float up to 40 V peak (28 V rms) from ground. When this value is exceeded the mains supply to the instrument is interrupted, the isolated grounding system is connected to supply grounding system, and an audible alarm is sounded. Applications include connecting the oscilloscope ground to the refer-ence rail instead of zero volts in ECL circuits to reduce probe loading, and reducing hum problems in low-level audio circuit measurements by avoiding earth loops.

An alternative approach to mains isolation is to disconnect the mains lead entirely. The Tektronix models 212, 213, 214 and 221 (see *Figure 6.4*) can operate either from a.c. mains or from internal rechargeable nickel-cadmium batteries. Of double-insulated impact-resistance plastic construction, these oscilloscopes when operated from their internal batteries can be floated at up to 700 V (d.c. + peak a.c.) above or below ground. Normal safety rules should always be observed when working with high voltages, especially where the ground lead of the probe is at other than earth potential. It is unwise to work on equipment with voltages in excess of 50 V if there is no other person in the same area.

Yet another alternative is signal voltage isolation. This is covered in the next chapter.

5

Using oscilloscopes

It seems superfluous to say that when using an oscilloscope to view a waveform, one should choose an instrument appropriate to the job in hand. Yet, as explained in the course of this chapter, besides the more obvious requirements ('does it have a bandwidth wide enough to display my signal faithfully?', 'is it sensitive enough to see the very small signal I wish to view?') there are quite a few other considerations that are a little less obvious. Some have already been pointed out, notably in Chapter 3, and others will become apparent in the course of this chapter. We shall also consider the case where there is no choice and one is faced with the task of trying to obtain some useful information about a waveform with an oscilloscope which is hardly adequate to the purpose.

Use of probes

Questions that people new to using oscilloscopes often ask are: 'Do I always need a probe? If not, how do I know when to use one and when not?' The first part of Chapter 4 should have provided a good deal of insight into this; if you are still puzzled it might be worth reading again. But for a short simple answer, the author's advice is always to use a 10:1 passive divider probe (correctly set up for the oscilloscope you are using) as a matter of habit. If, owing to the attendant attentuation factor of 10, the signal you wish to view gives insufficient vertical deflection even with the Y input setting at its most sensitive position, it will be necessary to consider whether it is possible to depart from your standard practice of using a probe.

For example, if you are using a metre or so of general-purpose audio screened lead to connect the signal to be viewed to the oscilloscope, the total capacitive loading on the circuit may well be several hundred pico-farads. This will be no consequence if looking at, say, the secondary voltage of a mains transformer, and generally acceptable for viewing the output of a hi-fi amplifier over the whole audio range. However, 200 pF has a reactance of 40 kΩ at 20 kHz, and you could well get a mislead-ing picture of a test waveform in one of the earlier high-impedance stages of the amplifier; worse, the phase shift caused by the additional capacitance could cause the amplifier to oscillate if the phase margin of its negative-feedback loop is rather sparse. Yet it is precisely in the earlier stages that you might want to avoid the attentuation of a passive divider probe.

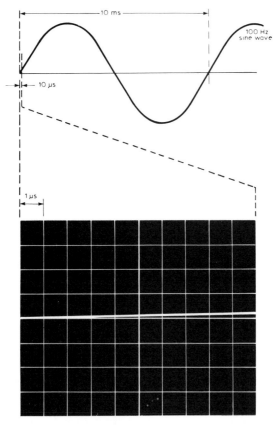

Figure 5.1 100 Hz sine wave displayed with 1 μs/div sweep speed. The 10 μs segment is not to scale, having been exaggerated for clarity. The display shows one-thousandth of a cycle, which would in practice be too dim to see

Two courses of action are open. For example, 75 Ω coaxial cable as used for television downleads has a capacitance of approximately 60 pF per metre, as against 150-180 pF for an audio screened lead. So if the connection can be made with a mere half-metre of lead, the total capacitive circuit loading including the oscilloscope's input capacitance can be kept down to about 60 pF. Alternatively, an active probe as described in Chapter 4 may provide the answer.

When investigating circuits operating at r.f. a passive divider probe is essential. Even with it, care must be taken if misleading results are to be avoided. For example, if the probe is connected across a tuned circuit, the extra 10-12 pF loading of the probe will change the resonant frequency to some extent. How much depends on how much capacity there is across the inductor to start with, but the probe's capacitance will be quite enough to upset the response of a conventional double-tuned i.f. stage, resulting in the wrong amplitude being displayed on the screen. The effect on an oscillator can be much more dire; not only will the connection of the probe cause a change in frequency, it will generally cause a reduction in amplitude and may well stop the oscillator altogether. The reason for this is that the input resistance 'looking into' the probe is 10 MΩ at zero frequency (d.c.), but falls steadily with increasing frequency. At several megahertz it may be down to a few hundred kilohms or even lower, imposing damping on the tuned circuit to which it is connected. In the case of an oscillator, of course, there will generally be several volts peak-to-peak of signal available, so if the oscilloscope is reasonably sensitive (i.e. has a 5 or 10 mV/div range) it will be possible to use a × 100 probe, with an input capacity of about 1 pF. If a × 10 probe is all that is available, much the same result can be achieved by connecting a 1.2 pF capacitor in series with the probe tip. The result will be a sort of × 100 probe that is not exactly calibrated and will not work at low frequencies. However, it will permit you to monitor oscillators, and tuned circuits generally, without affecting them unduly. In fact if it is only required to monitor the frequency and wave shape of the oscillator, the 1.2 pF capacitor can be dispensed with entirely and the tip of the × 10 probe simply held very close to, but not actually touching, the tuned circuit.

Trace finding

When using an oscilloscope to view waveforms, you will generally have some idea what to expect. Thus, if examining TTL logic gates operating at a clock frequency of 1 MHz, you would use a × 10 divider probe and set the scope Y input sensitivity to 0.1 or 0.2 V/div, d.c. coupled, giving 1 or 2 V/div sensitivity on the screen. The timebase speed would be set to, say, 1 μs/div.

However, it can happen that you do not know the appropriate settings, either because of a lack of information on the circuit under test, or because owing to a fault the waveform is not what you would expect. Let us suppose, then, that when the input signal is connected the trace disappears from the screen of the oscilloscope. The more expensive type of scope (and increasingly nowadays the cheaper models also) will have a trace-finder button; pressing this has the effect of restoring the trace to the screen regardless of the control settings, albeit in a defocused form. Its use is described in Chapter 3, but it should become unnecessary when you know how to drive an oscilloscope.

The commonest cause of a 'lost trace' is connecting a signal with a large d.c. component to the scope with the Y input d.c. coupled and the attenuator setting at too sensitive a position. So if you don't know what to expect, set the trace to the centre of the screen, set the Y input to a.c. coupled and the input attenuator to the least sensitive position— usually 20 or 50 V/div. It will then need a very large voltage to lose the trace, especially if using a 10:1 probe! In fact, with a.c. coupling, connecting a large d.c. voltage will move the trace up (or of course down if the voltage is negative), but the trace will then slowly return to the centre of the screen. This is so even if the Y input attentuator is at one of its more sensitive positions, although in this case it could be many seconds before the trace returns to the screen.

You can still lose the trace, even with the Y input a.c. coupled, if the input attentuator is at too sensitive a setting. Take, for example, a 1 kHz TTL square wave a.c. coupled to a scope set to a sensitivity of 5 mV/div: even using a 10:1 probe, the tops of the waveform will be off the top of the screen and the bottoms below the bottom of the screen. Although parts of the rising and falling edges will be on screen, they will be so rapid as to leave too dim a trace to be seen. If the scope has a trace-finder or locate button, pressing this will show two lines of dashes near the top and bottom of the screen, but if you always follow the sound practice of setting the Y input attenuator to the appropriate setting if known, or to the least sensitive setting if not known, you need never lose the trace in the first place.

The trace can also be lost through incorrect settings of the X time-base controls. Suppose, for example, you apply a 100 Hz sine wave to an oscilloscope, with suitable settings of the Y input controls but with the timebase speed set to 1 μs/div. When the timebase triggers, the sweep will be completed in 10 μs (assuming the screen has ten horizontal divisions). At the end of the sweep the spot will remain blanked for the next 9.99 ms until triggered by the next cycle; see *Figure 5.1*. With the trace blanked for 99.9 per cent of the time, it will be invisible, and on many cheaper scopes will remain so even if the intensity control is turned up. Only oscilloscopes with a high writing speed (see Chapter 7) will cope with this situation. The rule therefore is that if you do not know the frequency of the waveform you wish to examine, set the timebase speed to one of its slower settings, say 2 ms/div.

This leaves just one tricky case to watch out for, a narrow pulse occurring at a low repetition rate, say 100 ns wide at 100 pps (pulses per second). At 2 ms/div sweep speed the pulses will be too narrow to see and the trace will appear indistinguishable from the straight line produced by the auto brightline circuit. The test here is simply to switch to normal trigger, which disables the brightline circuit. Now the trace will only appear when the trigger level control is set to that part of its travel covered by the input pulse.

Practical examples

Having looked at the dos and don'ts of connecting a signal to an oscillo-scope, let us consider some practical measurement situations and see what they involve, starting with a simple case.

Example 1

Suppose you wish to look at the supply rail of a transistor amplifier to see just how much 100 Hz ripple there is. The amplifier probably has two supply rails, +24 and −24 V, so if you set the trace to the centre of the screen with the Y shift control and select 1 V/div setting, d.c. coupled, then with a 10:1 probe connected to one supply rail the trace will move up or down 2.4 divisions. (All the examples given assume a graticule format of ten horizontal by eight vertical divisions.) Most transistor hi-fi amplifiers nowadays operate in class B, which means that with the volume turned right down comparatively little current is drawn from the power supply. The 2.4 division vertical deflection will reveal that the supplies are indeed plus and minus 24 V, but the trace will almost certainly look exactly like a straight line.

If you wish to see the ripple, the Y input sensitivity must be increas-ed. If you increase it to 0.5 V/div (effectively 5 V/div in view of the probe) the trace will move off-screen. True, you can probably get it back with the aid of the Y shift control, but the ripple will still be too small to see and too small to trigger the timebase. The answer is of course to set the Y input to a.c. coupled, as this will block the d.c. voltage. You can now increase the Y input sensitivity as much as you like. With a 2 ms/div timebase speed you should see two complete cycles of ripple as in *Figure 5.2(a)*.

With a simple repetitive waveform like this, although the normal trig-ger mode can be used if required, triggering in the auto mode should be entirely satisfactory, provided there is enough Y gain available to give one division or so vertical deflection. If there isn't, this is a good example of a case where you can happily omit the 10:1 divider probe and simply use a screened lead. In the auto trigger mode the trigger circuit is a.c. coupled, so triggering will occur even for a fairly small waveform regardless of whether the trace is near the top of the screen or near the bottom, i.e. regardless of the d.c. component. It therefore

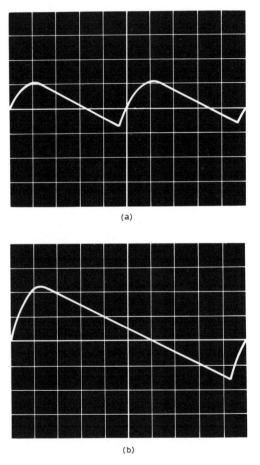

(a)

(b)

Figure 5.2 Ripple waveform across power-supply smoothing capacitor. (a) Full-wave or bridge circuit. (b) As (a) but with with faulty diode, or half-wave circuit. Horizontal scale 2 ms/div

follows that there is a frequency below which the auto trigger mode becomes progressively less sensitive. Depending on the particular oscilloscope design, this is usually in the range 10 to 50 Hz. When examining waveforms of too low a frequency for auto triggering, manual trigger, d.c. coupled, should be selected. Of course the Y input coupling must also be set to d.c., which means that if one wishes to examine a very small amplitude low-frequency signal riding on a large d.c. component one has problems. However, referring to *Figure 5.2*, auto trigger is ideal for the purpose. In the auto mode, in the absence of an input waveform to trigger the timebase, circuitry internal to the oscilloscope will (on virtually all makes and models) cause the timebase to run repetitively. This results in a 'brightline' on the screen, avoiding a 'lost trace' in the absence of a Y input.

If the waveform is like *Figure 5.2(b)* this reveals straight away that one of the diodes in the bridge rectifier is faulty. Without an oscilloscope you might by deduction have suspected this, though to confirm it, you would have had to disconnect components; but with an oscilloscope the diagnosis is easy.

Example 2

Now let us consider a slightly more complex case. Suppose you wish to examine the waveforms produced by a TTL decade divider such as an SN7490 running at a fairly high clock rate. The various waveforms are as shown in *Figure 5.3*. The input waveform and the output waveforms of the first, third and fourth stages are all simple repetitive waveforms (although only the input and the first-stage output waveforms are square waves in the sense of having a unity mark/space ratio); so with any appropriate timebase speed there will be no problem with triggering, either using normal trigger or in the auto mode.

However, the Q_B output from the second divider stage is a little more tricky. Using internal triggering as we have been up to now, bear in mind that, depending on the clock frequency and the timebase speed, if one trace commences at point 1 (positive-going trigger selected) it could terminate at point 2 (thus displaying rather more than one complete cycle of the Q_B waveform) and trigger again at point 3. The section of waveform displayed on this next scan would therefore be displaced horizontally relative to the previous scan, with the result shown in *Figure 5.4*.

There are a number of ways round this problem, depending on the facilities available on the scope in use. Consider first a very basic scope, and assume also that you cannot change the waveform's clock frequency. If the scope has a continuously variable timebase control, you can use this to set a slightly slower speed so that the sweep terminates just after point 3. The next sweep will then commence at the next point 1, and an unambiguous display will result. There are two snags to this solution, however. First, the timebase is now uncalibrated, which is inconvenient; second, the majority of inexpensive scopes do not possess a variable timebase speed control. But virtually every scope, even the cheapest, has an external trigger input, so the straightforward solution is to apply the Q_D waveform to this. On a dual-trace scope, one can alternatively display the Q_B waveform on the Y1 trace and the Q_D waveform on Y2, with internal triggering selected from the Y2 channel.

If it is particularly desired to display the clock and Q_B waveforms simultaneously (for example when measuring the propagation delay through the first two stages of the counter), applying the Q_D waveform to the external trigger input is again an appropriate course of action. Of course it involves making an extra connection between the oscilloscope and the circuit under test, but this minor inconvenience can be avoided if the scope has a trigger holdoff control. Trigger holdoff was mentioned briefly in Chapter 3 and is a very useful facility in this situation. Normally an oscilloscope trace is available to be retriggered as soon as the flyback is completed, and this is the case when the holdoff control is in the normal (fully anticlockwise) position. As it is rotated further clockwise, there is an increasing delay between the end of flyback and the time when the trace can next be triggered. The maximum delay or holdoff is generally several times the sweep time, depending on the make of oscilloscope. With this control it is thus possible to obtain a stable display using internal triggering from the Q_B waveform with the timebase speed at a calibrated setting regardless of the clock frequency.

Manual triggering

Bearing in mind the above points, manual triggering from waveforms either digital or sinusoidal is straightforward: simply select manual (normal) trigger, positive or negative trigger polarity as required, and adjust the trigger level control to cause the trace to commence at the desired level on the chosen edge of the waveform. Triggering from, say, the positive-going edge of a sine wave will then be possible from a point slightly above the negative peak, right up almost to the positive peak. The exception is when examining fairly low frequencies with the l.f. reject trigger facility in use, or fairly high frequencies with h.f. reject in use. These controls cause a progressive decrease in trigger sensitivity at low and high frequencies respectively; their use is covered in Chapter 3. Besides decreasing the effect of unwanted low- or high-frequency components on triggering by the wanted waveform, these controls have an effect on the triggering that is worth noting.

Suppose you are using h.f. reject on an oscilloscope where this mode rolls off the high-frequency response of the trigger channel above 50 kHz, in order to obtain a stable display of a 50 kHz sine wave that has superimposed some low-level narrow spikes of an unrelated frequency. The trigger circuit will now reject the spikes and respond only to the wanted 50 kHz sine wave, which will thus be cleanly locked although the spikes may be visible running through. If very narrow, they may well be quite invisible at the timebase speed used to view the

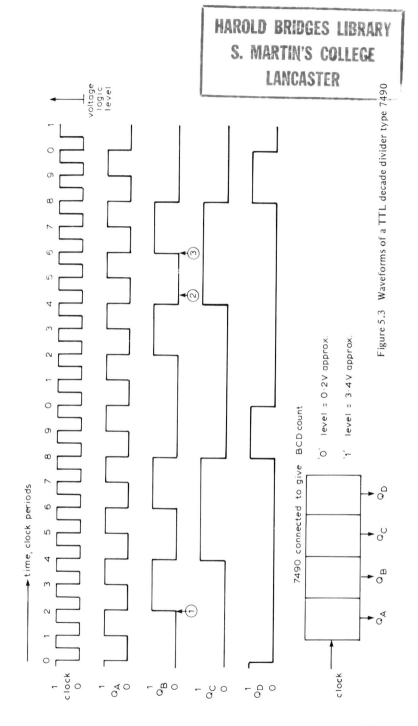

Figure 5.3 Waveforms of a TTL decade divider type 7490

7490 connected to give BCD count

'0' level = 0·2V approx.

'1' level = 3·4V approx.

49

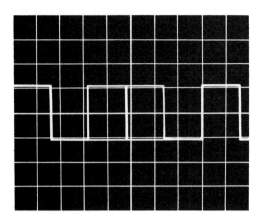

Figure 5.4 Q_B trace overlapped owing to triggering at points 1 and 3 (see Figure 5.3) alternately

wanted 50 kHz signal, yet without the h.f. reject facility they could have made it quite impossible to obtain a locked picture of the wanted signal. Now the trigger circuit will respond to the 50 kHz wanted waveform, although its response will be 3 dB down, i.e. the smallest 50 kHz signal it will lock on is about 40 per cent larger than at lower frequencies. In addition, there will be a corresponding $45°$ phase lag in the trigger channel. The significance of this is that if you have selected manual trigger, positive-going, the trigger level control will no longer enable you to trigger at any desired level on the positive-going flank of the sine wave. Instead, the trigger level control will initiate the sweep anywhere from one quarter of the way up the positive-going flank to one quarter of the way down the following negative-going flank. At frequencies higher than 50 kHz this effect will become more pronounced.

A similar effect will be noted when triggering from a waveform near the l.f. reject cut-off frequency with l.f. reject selected, except that in this case there will be a phase advance, so the trigger range will be advanced by up to a quarter of a cycle rather than retarded as in the h.f. reject case.

Use of dual-trace scopes

It is frequently convenient, and indeed essential to be able to view two waveforms simultaneously. This facility has been available in scopes near the top end of the market since before the Second World War. However, for many years the means of achieving this was to use a cathode-ray tube with two beams, each with its own Y deflection system but sharing common X deflection plates. The two beams could be produced by two independent electron gun assemblies, or by a single gun and a 'splitter plate' to slice the beam in half. With an oscilloscope using this type of dual-trace operation, known as a 'dual-beam' oscilloscope, one could always be certain that the two waveforms viewed really were in the time (or phase) relationship shown, since both beams were deflected simultaneously by the common X timebase waveform.

Stemming from the advances in electronic circuitry made during the Second World War in connection with radar displays, it became possible to use a single-beam cathode-ray tube to display two (or more) traces. The resultant simplification of the cathode-ray tube enabled designers to concentrate on producing tubes with higher writing speed and greater deflection sensitivity, especially in the Y axis. The importance of this to the evolution of oscilloscopes with better performance generally and wider bandwidth in particular is covered in more detail in Chapters 7 and 8, but the return to single-beam tubes has particular significance when using a dual-trace scope to examine the relative timing of two waveforms.

To display two traces the single beam must somehow be shared between the Y1 and Y2 traces. Dual-trace oscilloscopes almost invariably offer a front-panel selectable choice of alternate and chopped modes as well as, usually, Y1 only and Y2 only, as described in Chapter 2. Likewise there is usually also a choice of triggering source: Y1, Y2, mixed or external. The chopped mode of display writes the two traces during each single sweep. It achieves this by writing a very short portion of the Y1 trace, then a portion of the Y2 trace, and so on alternately. Each trace therefore consists of a series of dashes, but when displaying low-frequency signals the dashes merge to provide the appearance of two continuous traces. Compared to the alternate mode, where first a complete Y1 trace is written and then a second sweep writes the Y2 trace, the chopped mode results in the absence of flicker down to half the sweep repetition rate at which flicker appears in the alternate mode. On the other hand, when displaying frequencies of a few kilohertz or higher, the dotted-line structure of the two traces in the chopped mode may become apparent as the chopping rate is generally between 100 kHz and 1 MHz. Thus the chopped mode is most suitable for low-frequency signals and the alternate mode for higher-frequency signals, say a few kilohertz upwards.

For 'single-shot' operation, for example when photographing the screen to record what happens at two points in a circuit following the operation of a pushbutton, the chopped mode is obviously appropriate, since in the alternate mode only one of the two traces will appear. If the signals to be observed are such as to require a timebase speed too high for the chopped mode to be useful, then it is impossible with a dual-trace real time scope to observe both channels on a single-shot basis, a limitation that does not apply to the older true dual-beam oscilloscopes (see page vi). For this reason, true dual-beam oscilloscopes are still available. The Gould Advance OS260, now discontinued, was probably the last small budget-priced true dual-beam oscilloscope. True dual-beam oscilloscopes are still available in the higher price range, for example the Tektronix 7844 (400 MHz bandwidth) and 5113 (bistable storage).

It was stated earlier that in the chopped mode, for signals above a few kilohertz, the dotted-line structure of the traces *may* become apparent. However, in general there will be no fixed frequency relation between the signal being viewed (to which the trace repetition fre-

Figure 5.5 The CCD camera of the Tektronix DCS, together with its connecting leads and interface board. The latter fits in an expansion slot of an IBM personal computer, to form the complete Digitizing Camera System

quency is locked) and the chopping frequency; so the missing portions of the Y1 trace on one sweep, where the beam is writing part of the Y2 trace, will be partly or completely filled in on the next sweep, and so on. Given an a.f. (audio frequency) oscillator with a good slow-motion dial drive though, it is quite an easy matter to adjust its output at around 10 kHz to a subharmonic of the chopping frequency. As the right frequency is approached, the dashes of which each trace is composed can be seen running across the trace, and with a little care (and a stable oscillator) they can be made stationary. The slightest mistuning should cause them to run through. On some oscilloscopes they will stay locked with very slight mistuning, suddenly commencing to run through with further mistuning, but this is a sign of poor design or construction, resulting in crosstalk between the chopping-frequency generator and the trigger circuitry. This will not be the case on most well-known makes of dual-trace oscilloscope; when the signal and chopping frequency are not related, as is usually the case in practice, the chopped mode can be used for repetitive waveforms at frequencies right up to and beyond the chopping frequency, though there is little point in so doing.

The choice of trigger source is very important when working with a dual-trace oscilloscope. As mentioned earlier when discussing the waveforms encountered in a decade-divider stage, if the frequencies being displayed on the Y1 and Y2 traces are different but related, one should trigger from the lower frequency, whether it be displayed on the Y1 or Y2 trace. Many dual-trace oscilloscopes have a 'mixed trigger' facility; this means that when used in the alternate mode with internal triggering, the sweep will be triggered from the Y1 signal when displaying the Y1 trace, but on the next sweep will display the Y2 trace triggered from the Y2 channel. Consequently both channels will be perfectly synchronised and the traces will appear to have a fixed stable relationship. In fact, the signals being displayed on the two channels could have totally unrelated frequencies, as would be apparent if triggering from Y1 were selected, in which case the Y2 trace would not be synchronised, and vice versa.

In the mixed triggering mode, the oscilloscope is simply equivalent to two entirely separate single-channel scopes, each internally triggered from its own signal. Nevertheless mixed triggering can be very useful for keeping an eye on two unrelated waveforms simultaneously, provided this fact is borne firmly in mind. Care is needed even when the two frequencies are closely related or identical. Mixed triggering will show the $0°$ reference output and the $90°$ quadrature output of a quadrature oscillator as being in phase, whereas triggering from the reference input on the Y1 channel will show the correct $90°$ phase difference between the two sine waves. The moral is to use mixed triggering only when it is specifically required, and to regard the selection of the appropriate triggering arrangements as an essential part of setting up a dual-trace scope.

Many dual-trace oscilloscopes provide the option of displaying a single trace which represents the sum of the voltages applied to the Y1 and Y2 inputs. In addition, it is possible to invert one of the traces, say Y2, so that positive going inputs deflect the trace *downwards* and negative inputs *upwards*. It is thus possible to display Y1–Y2, i.e. the *difference* between the two inputs. This will result in no deflection of the trace if the *same* signal is applied to both Y inputs – provided that they are set to the same volts/division setting (and both variable controls, if provided, are at the calibrated position). Thus the oscilloscope will only respond to the difference between the two inputs, just what is wanted for examining two wire signals that are balanced about ground.

This property of ignoring or rejecting identical signal components at the two inputs is called 'common mode rejection' or 'input balance'. The unwanted 'push-push' component that is rejected is called 'common mode noise', 'longitudinal noise' or 'noise to ground', whilst the push-pull signal is called the 'transverse', 'metallic' or 'normal mode' signal. Two-wire balanced transmission systems are widely used, e.g. for transducer signals in factory process control systems, as twisted pairs in multi-pair telephone cables and for the two-wire overhead 'subscriber's loop' connecting the domestic telephone to the nearest telegraph pole.

The Y1–Y2 mode will typically provide a 26 dB CMRR (common mode rejection ratio), meaning that the sensitivity to undesired common mode signals, e.g. 50 Hz mains hum, is only one twentieth of the sensitivity to the wanted transverse signal. This is only a modest degree of input balance compared with special oscilloscopes and other instruments specifically designed for working on balanced transmission systems. However, balanced systems are generally used only up to a few hundred kilohertz at most, and instruments specifically designed for such use are correspondingly limited in bandwidth. Note that if 10:1 passive divider probes are in use, the 20:1 CMRR may be degraded to 10:1 (20 dB) or less, owing to within-tolerance differences in the exact division ratios of the two probes. With or without probes, the CMRR can be optimized by connecting both inputs to the same signal source and adjusting one or other variable gain control to trim down the gain of one channel to match that of the other. With care, up to 100:1 CMRR (40 dB balance) or more can be obtained for signals up to a few hundred kHz, but this will not usually be maintained over the full bandwidth of the oscilloscope. To maintain this increased CMRR, re-adjustment will also be necessary if the two Y input volts/division switches are set to another (common) setting.

When using an oscilloscope's Y1–Y2 mode for balanced measurements, beware of a potentially severe limitation. If the unwanted common mode signal (e.g. mains hum) is much larger than the desired signal, it can overload the Y amplifiers, resulting in a distorted and inaccurate display. This problem can be avoided by using a purpose-designed differential probe. In the Tektronix P6046 Differential Probe and Amplifier Unit, the differential signal processing takes place in the probe itself, the amplifier producing a single-ended 50 Ω output suitable for connection to any oscilloscope Y input channel. The P6046 provides 10,000:1 CMRR at 50 kHz and no less than 1000:1 even at 50 MHz, while common mode signals up to ±5 V peak to peak (±50 V with the clip-on X10 attenuator) can be handled without overload, even when examining millivolt level signals.

In power engineering it is often necessary to examine small signals in the presence of very large common mode voltages, for example when checking that a silicon controlled rectifier's gate to cathode voltage excursion is within permitted ratings, in a motor control or inverter circuit. The Tektronix A6902B Voltage Isolator uses a combination of transformer – and opto-coupling to provide up to ±3000 V (d.c. + peak a.c.) isolation from ground for each of two input channels. Designed for use with any two channel oscilloscope, the A6902B permits simultaneous observation of signals at two different points in the same circuit, or signals in two different circuits without respect to common lead voltages. The two channels can also be combined to function as an input to a differential amplifier, for floating differential measurements.

Lissajous figures

On most oscilloscopes one can turn off the internal timebase and apply an external signal to the X amplifier. This usually has a more limited bandwidth than the Y amplifier, but the facility can be extremely useful. One particular application, the Lissajous figure, is important enought to describe in some detail.

Suppose that the same sine-wave signal is connected to the Y and X inputs, and the controls are adjusted so that the peak-to-peak vertical deflection equals the peak-to-peak horizontal deflection. The result will be a straight-line trace on the screen, sloping at 45° from bottom left to top right. The line will be brighter at each end than in the middle, since a sine wave is changing most rapidly as it passes through zero voltage.

Now suppose that, instead of applying the same sine wave at both X and Y input, one of the sine waves is inverted, i.e. 180° out of phase with the other, such as would be obtained from the ends of a mains-transformer secondary winding whose centre tap is earthed. The result is a straight line again, but this time sloping from top left to bottom right. If the two sine waves are in quadrature, i.e. have a phase difference of 90°, the result is a circle; if they are almost in phase, but not quite, the straight line becomes a long, thin ellipse.

If, instead, we apply two different sine waves differing in frequency by 1 Hz, then once every second the sloping line becomes an ellipse, fattening out to become a circle, thinning through an ellipse sloping the other way to a straight line at right angles to the first, and then waxing and waning again back to where we started. If the frequency difference is larger, say 10 kHz and 10.1 kHz, the rolling of the Lissajous figure will be too fast to follow and we will see a square of light, fainter in the middle and brighter at the edges and corners. We have in fact a method of comparing the frequencies of two sine waves, since if we adjust the frequency of one oscillator until the Lissajous figure becomes stationary, be it line, circle or ellipse, then its frequency must equal that of the other oscillator.

The line-ellipse-circle representing unit frequency ratio is the simplest kind of Lissajous figure. The traces in *Figure 5.6* show frequency ratios of 3:1 (Y frequency: X frequency) and 3:2. Traces *(b)* and *(c)* show the same 3:2 ratio, but with different phases between the two signals. It is

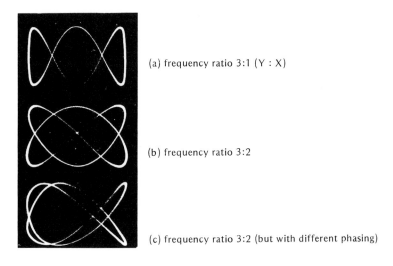

(a) frequency ratio 3:1 (Y : X)

(b) frequency ratio 3:2

(c) frequency ratio 3:2 (but with different phasing)

Figure 5.6   Lissajous figures (courtesy AEG Telefunken)

quite straightforward to tell the frequency ratio from the figure by counting the number of times the figure cuts a vertical and a horizontal line drawn across it. Thus in *Figure 5.6(a)* a vertical line down the centre of the figure cuts it in two places, and a line across the figure near the top (always avoid taking the line through a point where the branches of the figure cross) cuts it at six points. Therefore the ratio of the X and Y frequencies is 1:3.

The Lissajous figures usually shown are those produced by sine waves, but square waves will also produce Lissajous figures of a sort. For example, equal-frequency square waves in phase with each other simply produce two dots of light, one at the bottom left, the other at the top right, though they will be joined by a very faint line if the rise and fall times are an appreciable fraction of a cycle. It is interesting, instructive and not difficult to work out what happens if the square waves are of the same frequency but in quadrature, and if they differ in frequency by 1 Hz. It is not of merely academic interest either. The complex Lissajous figure produced by two unrelated multilevel square-waves was used as a diagnostic display in the Hewlett Packard 1600S Logic Analyser.

Use of Lissajous figures

It might seem that nowadays the use of Lissajous figures for comparing frequencies is 'straight out of the Ark'—why not simply use a frequency counter? There are several cases where the use of Lissajous figures can provide more information and provide it faster.

Suppose, for example, one had a precision 1 MHz frequency standard consisting of an oscillator controlled by an ovened crystal. One could check its frequency with a digital frequency meter if the latter's internal reference were accurate enough or could be independently checked. In the UK, one could check by counting the carrier frequency of the BBC's Droitwich transmitter, whose carrier is maintained to an accuracy of one part in 10^{11}. In fact, 'off-air frequency standards' are available commercially; these receive the Droitwich transmission, strip off the amplitude modulation and supply a 1 MHz output locked to the carrier. However, even a 10-second gate time will only allow a 1 MHz frequency to be checked to an accuracy of ± 1 count in 10^7, which makes checking the frequency meter and adjusting the 1 MHz oscillator a tedious business. Even then the accuracy finally achieved will fall far short of that available from the Droitwich carrier.

Suppose now that a Droitwich-derived 1 MHz sine wave and the crystal oscillator under test are displayed as a Lissajous figure; the effect of adjusting the crystal oscillator can be observed immediately and continuously. A frequency difference of as little as one-hundredth of a hertz can be noticed during an observation time of a second or so, whereas a counter would still have an uncertainty of plus or minus one-hundredth of a hertz after a 100-second gate period.

Lissajous figures can also provide information about the stability and spectral purity of an oscillator. For example, if two conventional signal generators (using LC oscillators) are both set to 100 kHz the resulting Lissajous display should be stable, giving a clean line and a round circle as the inevitable small frequency difference causes the figure to cycle slowly through its series of patterns. If now a Wien-bridge type of RC oscillator is substituted for one of the signal generators, the poorer frequency and phase stability of this type of oscillator will be immediately apparent. The circle, instead of being round, may show minor dents and the figure will wobble, rather like a jelly being carried on a plate. This is evidence of very low frequency noise f.m. sidebands, which it would be very difficult to resolve with even the most sophisticated spectrum analyser. It is in fact this phase instability that would

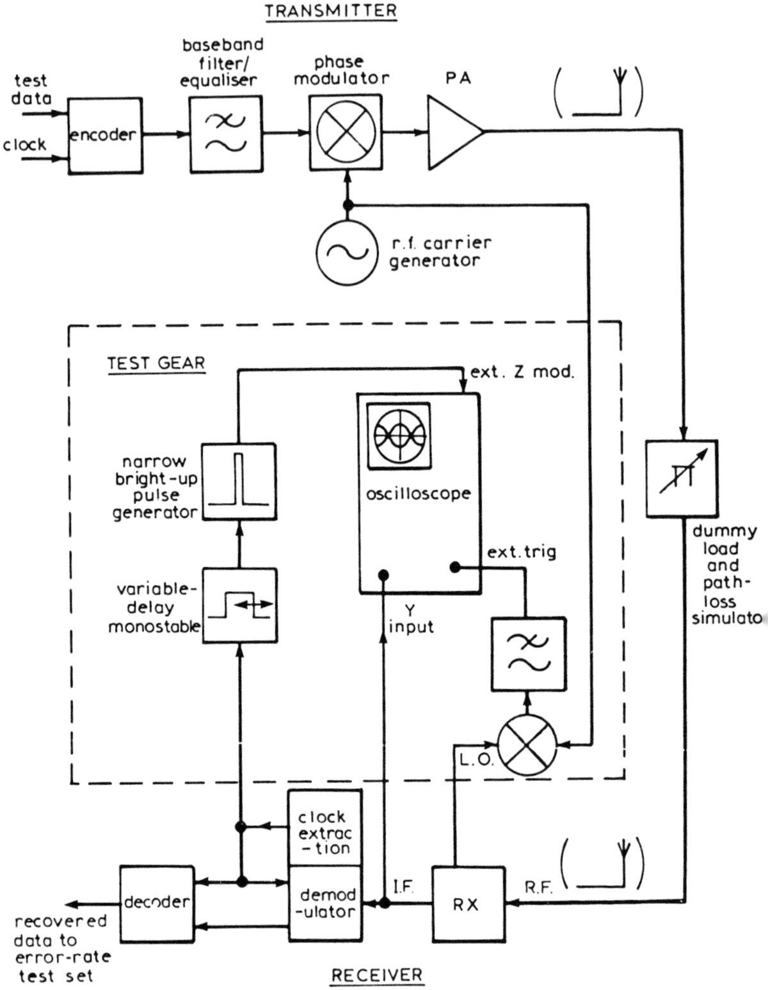

Figure 5.7 Block diagram of digital phase-modulation radio data link on test (simplified)

set the limit to how accurately one could compare the two 1 MHz signals in the previous example.

Z axis input

A useful feature of many oscilloscopes is a 'Z axis' input. In Cartesian coordinates the Z axis is the third dimension at right angles to the X

58

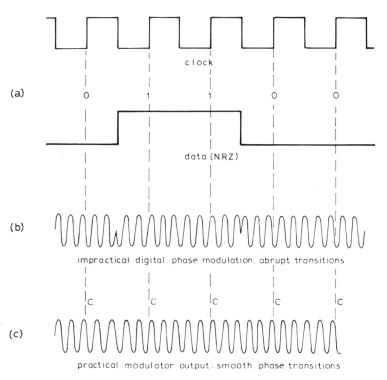

(a)

clock

0   1   1   0   0

data (NRZ)

(b)

impractical digital phase modulation: abrupt transitions

c   c   c   c   c

(c)

practical modulator output: smooth phase transitions

Figure 5.8 (a) Clock and typical data stream of data link shown in Figure 5.7. (b), (c) Modulated r.f. output waveforms; note that both have the same phase at clock times 'c'. For clarity the r.f. is shown as exactly five times the clock frequency; in practice it would be many thousands of times and with no exact relation

and Y axes, and therefore the same as the direction of the electron beam in the cathode-ray tube when the spot is at the centre of the screen. The author has always considered the expression 'Z axis modulation' somewhat unclear and inappropriate, but it is firmly entrenched in the language of oscillography.

With no connection made to the Z axis input, the oscilloscope works normally with the trace brightness controlled by the intensity control, also affected of course by the sweep repetition rate and timebase speed as explained earlier. Applying a varying voltage to the Z axis input alters the brightness of the trace in sympathy. Some oscilloscopes have d.c. coupling of the Z axis input, but a.c. coupling is cheaper and therefore more common, while positive-going voltages result in a decrease of brightness if, as is commonly the case, the Z axis input is coupled to the cathode of the c.r.t.

The facility is of use in a variety of ways, one interesting example being the display of 'eye diagrams'. These are a way of examining the degradation due to imperfections of the modems in a digital phase-modulated communications link—*Figure 5.7*. Normally the receiver for such a transmission has a 'clock' or timing recovery circuit; displaying the received waveform with the scope triggered from the recovered timing results in a very jumbled picture. Bandwidth is a scarce and hence expensive commodity, and the sudden changes of phase shown in *Figure 5.8(b)* imply the presence of wide sidebands. The transmitted modulated carrier is therefore first filtered and limited to produce a smoothly changing phase—*Figure 5.8(c)*.

To display an eye diagram, the recovered timing is used to generate a narrow pulse occurring at the clock edge or data-stable time. This is used to bright-up the oscilloscope trace, which runs continuously, triggered from the unmodulated carrier. As the trace is invisible except during the bright-up pulse, i.e. at the sampling instant of the receiving modem, the phase of the received signal will be in one of two possible positions. The resultant picture is called an eye diagram. In *Figure 5.9* the 'open' eye such as might be obtained with a well set-up system, indicates little distortion; the nearly 'closed' eye shows a poor system with intersymbol interference. (*Note:* the above is a much simplified description of digital phase modulation; practical systems use differential rather than absolute phase modulation, often of the symmetrical type rather than the unsymmetrical type described.)

The oscilloscope in servicing

Several of the facilities of a good oscilloscope have been discussed above in connection with specific applications. In the rest of this chapter we look at other particular areas of use for an oscilloscope. First, TV servicing is considered briefly; for a more extensive treatment of the topic reference should be made to one of the many excellent books available dealing specifically with this subject.

It is important to pay due regard to safety when working on any type of mains-operated equipment. This is doubly true when working on TV sets, as some of them do not have the circuitry and chassis isolated from the mains. The circuitry of the ubiquitous 12 in black and white set is designed to run from 12 V d.c., in order to permit 'portable' operation from a car battery when required. For mains operation, a step-down transformer, rectifier and smoothing supply the required 12 V d.c. Thus only the transformer primary is at mains potential, the rest of the set being isolated. Larger mains only colour TVs may have a type of switchmode power supply providing full mains isolation, but this is by no means invariably so. To avoid drawing a d.c. component from the mains (this was quite normal in the days of valved TV sets), non-isolated sets use a full-wave rectifier; as a result the set's circuitry and chassis can be at approximately half mains voltage.

The only safe way to proceed when working on a TV chassis is to run it from a mains isolating transformer of suitable rating. A 500 VA transformer should be more than adequate. The television set's chassis should be firmly earthed, as is the case of the oscilloscope. Even then, one must be very wary of the high voltages present in the line deflection and e.h.t. sections of the receiver.

No one should work on a TV set without adequate knowledge and expertise. Even apart from the safety aspect, many faults will prove difficult or impossible to rectify without the full servicing manual for the particular model. Occasional loss of colour, for example, can be due to a variety of causes, and adjusting controls in the wrong sequence can easily give you permanent loss of colour!

The most convenient kind of scope for TV servicing has built-in line and frame synchronising separator circuits, e.g. the Farnell DT12-5 featured in Chapter 2. These are useful when examining the operation of line and frame deflection circuits respectively, particularly when the set is receiving live programme material. The TV sync circuits enable the scope to be triggered stably from the output of the video detector. However, it should be possible to trigger any good oscilloscope from the line sync component of the video waveform by selecting normal trigger, positive or negative slope as required, and adjusting the level control to trigger on the tips of the video, i.e. the sync pulses. Problems may be encountered in the case of a cheaper black and white set (with mean level a.g.c. applied to the vision i.f.) on programme material, as the sync peak amplitude will change with scene changes, the video being a.c. coupled. This problem is easily solved by using instead the signal from a grey-scale or colour-bar generator.

Servicing hi-fi equipment is a less complex task than TV servicing. An a.f. signal generator and a scope should enable sufficient testing for all practical purposes to be carried out. Dummy load resistors, of a suitable rating, to replace the speakers during full power testing can be described as a necessity rather than a luxury. A sine-wave test signal can be followed through the various stages and any gross distortion observed and pinpointed to the offending stage. In addition to clipping the signal on positive or negative peaks (usually a sign of a faulty bias-resistor network), a wideband oscilloscope may reveal that the amplifier becomes unstable with bursts of oscillation on one or both peaks of the waveform at full drive, while behaving normally at lower drive levels. On live programme material this can give rise to a nasty tearing noise appearing in loud passages only. Quite apart from these extreme forms of distortion, an amplifier (or much less often a pre-amplifier) may exhibit one

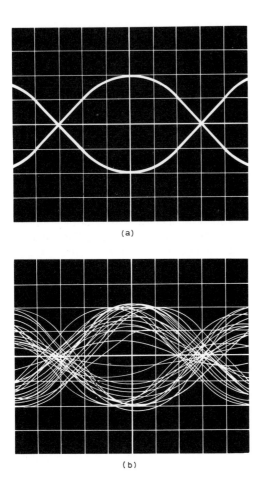

(a)

(b)

Figure 5.9 (a) Two-level digital phase-modulated signal showing well-set-up system with no intersymbol interference. (b) Poor system with bad intersymbol interference

or two per cent distortion, usually more noticeable at higher volume levels. It is very difficult to detect even several per cent of third-harmonic distortion simply by examining the output waveform on an oscilloscope, but the diagnosis is much easier if it is possible to display the undistorted input test sine wave on the other trace of the scope for comparison. Unlike third-harmonic (and other higher odd-order) distortion components, second and other even-order harmonic distortion affects the positive and negative half-cycles of the waveform differently, usually making one flatter and the other more peaky. Consequently, with care even 1 per cent of second-harmonic distortion can be readily seen by examining the trace. Of course, even 1 per cent of distortion completely disqualifies an amplifier from any pretence to the title 'hi-fi', but it is surprising how many of the less expensive amplifiers on the market, especially those forming part of a cheap packaged 'music centre', do little if any better than this, particularly at the extreme bass and treble ends of the audio range at full power output.

To measure the distortion in a 'real' hi-fi amplifier a scope will not suffice. A total harmonic distortion (t.h.d.) meter is required, to remove the original sine wave from the amplifier output and measure the relative amplitude of the residual signal. This consists of harmonics, noise and very often 100 Hz hum from the power supply. Many t.h.d. meters make the residual signal available for examination on an oscilloscope, which can be very informative. For example, once the fundamental is removed it is easy to see whether second or third harmonic predominates, while the presence of little pips of alternate positive- and negative-going polarity indicates 'crossover' distortion in a class B amplifier. Often, also, class B amplifiers show considerable 100 Hz ripple in the residual at full power output, due to penny-pinching in the size of the smoothing capacitors of the power supply. At low volume the class B output stage takes little current so the hum is not noticeable, while at full output the designer relies on the loud programme content to mask the hum.

Provided it has sufficient bandwidth to cope with the signal, an oscilloscope can be very useful when developing or troubleshooting radio-frequency circuits. The main point to watch for here is the effect of the loading imposed on the r.f. circuit. Even using a good 10:1 divider probe, the mere act of looking at an r.f. circuit can detune it or cause it to oscillate. This has already been covered in Chapter 4, which contains suggestions for coping with the problem, so no more will be said on that topic here.

Bandwidth

The bandwidth quoted for an oscilloscope generally refers to the frequency at which the amplitude response has fallen by 3 dB. Remember therefore that if examining a 25 MHz sine wave with an oscilloscope with a quoted 25 MHz bandwidth, the trace will show only 71 per cent of the true amplitude of the signal. Furthermore, the waveform being observed may be rather severely distorted if not a sine wave, since the harmonics have frequencies of 50, 75 MHz etc., and the oscilloscope's response at these frequencies will be very low indeed. Even an ideal 25 MHz square wave will look tolerably like a sine wave on a 25 MHz scope! Actually the situation is a little more complicated than this. The rated bandwidth of the highest-quality oscilloscopes is quoted as that frequency at which the full screen response has fallen by 3 dB. Other manufacturers quote the −3 dB frequency for half-screen-height deflection, while one North American manufacturer who shall remain nameless quotes the bandwidth at the −6 dB (50 per cent response) point for quarter-screen-height deflection. Let the buyer beware.

Figure 5.10 Good examples of modern high-performance oscilloscopes, the Solartron Schlumberger 5220 and 5224 both provide 100 MHz bandwidth. The 5220 has two full Y input channels plus a third with fixed sensitivity; the 5224 has four full Y input channels. Top of the range 5229 (Figure 1.6) boasts 500 MHz bandwidth (courtesy Solartron Instruments)

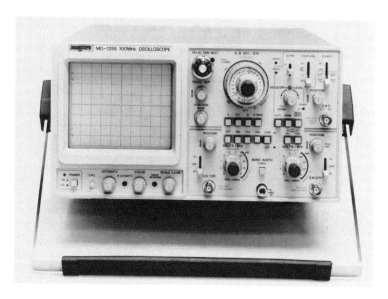

Figure 5.11   The three channel 100 MHz Meguro model MO-1255 displays up to 8 traces. The A and B timebases cover speeds from 0.5 s/div to 20 ns/div, or 2 ns/div with the X10 magnifier in use (courtesy Fieldtech Heathrow Ltd)

The screen height (full screen, half screen or whatever) at which the bandwidth is quoted is important, as the amplifiers driving the Y deflection plates are nowadays usually slew-rate limited; this is explained in more detail in Chapter 8. It means, however, that provided one contents oneself with considerably less than full-screen deflection, the bandwidth of the oscilloscope is usually greater than the quoted figure. Unfortunately, as the frequency rises, so does the amplitude required to operate the trigger circuitry. Thus when trying to observe a signal that is beyond the full-screen bandwidth of the oscilloscope, the amount of extra bandwidth to be had by reducing the displayed amplitude is limited to the point at which the trigger circuit ceases to function. If one is simply trying to see whether there is a signal of any sort there (for example, trying to find out whether the local oscillator of a Band II f.m. receiver is working, and on approximately the right frequency, using a 40 MHz oscilloscope), then the Y input sensitivity can be increased, if need be right up to maximum. No quantitative amplitude measurement will be possible of course, but as a qualitative indication of whether the oscillator is functioning or not, useful information has been obtained.

Just how far beyond the maker's nominal bandwidth an oscilloscope can still provide useful information depends not only on the frequency at which the trigger circuit gives up the ghost, but on the design philosophy of the Y amplifier. If the designer was aiming at maximum bandwidth, the frequency response may have been propped up at the top end with numerous bits of compensation and peaking circuits. In that case, at frequencies beyond the design maximum, the Y amplifier response may fall away very rapidly indeed. But the designer may have been aiming instead at a low rise time for pulse and square-wave signals, coupled with no ringing and only one per cent or so of overshoot when displaying a signal with a rise time much less than that of the oscilloscope. In this case the fall-off of the frequency response beyond the nominal bandwidth will be more gradual.

With the ever-increasing importance of digital circuitry, most modern oscilloscopes will exhibit this more desirable form of high-frequency response. The result is that an oscilloscope with a 5 MHz, −3 dB full-screen bandwidth may be able (given sensitive enough trigger circuitry) to display a 10 MHz square wave at one or two divisions vertical deflection. Of course it will not appear quite like a square wave, but on the other hand the flattening of the peaks will make it clear that it is not a sine wave. The oscilloscope is doing its best to tell you that the waveform is square! Thus, used with intelligence and understanding, an oscilloscope can provide considerable useful information, even if only of a qualitative nature, when handling signals well beyond its nominal capabilities.

Finally, *Figure 5.10* shows more examples of modern high-performance real-time oscilloscopes with all the facilities mentioned in this chapter.

6

Oscilloscopes for special purposes

It would be very difficult, indeed quite impossible, to design an oscilloscope suitable for all the very wide range of uses to which this most versatile of electronic instruments is put. Consequently there is and always will be a wide variety of oscilloscopes, each aimed primarily at its own peculiar field of application.

Of course the mainframe plus plug-in approach permits one oscilloscope (plus a cupboard full of plug-ins) to cover a wide variety of uses, but this format is confined to medium and large oscilloscopes. Even then there is usually a difficult choice to be made. Either the mainframe will have a storage cathode-ray tube, which makes it more expensive than one with a conventional c.r.t., or it will have the latter, in which case it is generally capable of a greater bandwidth than a storage mainframe, especially one of comparable cost. But more of this later when we consider plug-in and storage scopes in particular. First, let us consider the smaller, simpler specialised instruments.

Small portable scopes

Being such versatile instruments, oscilloscopes often get used in inaccessible places, down a hole in the ground for example, or at the top of a pole. Here a small light instrument, powered from internal batteries, has obvious advantages. *Figures 6.1* to *6.4* show a selection of such instruments, some powered from internal primary ('dry') batteries and some from internal secondary (rechargeable) batteries. Often the latter variety incorporates a mains-powered battery charger, and depending on the model it may also be possible when mains is available to use it while simultaneously recharging the batteries for later portable use.

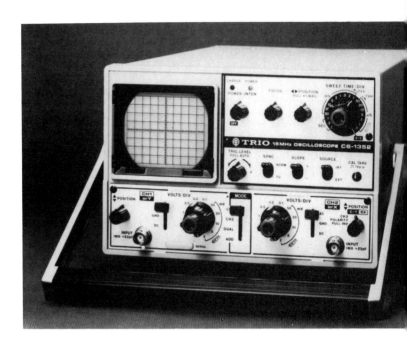

Figure 6.1 Trio 15 MHz CS1352 operates from a.c. mains, external 12 V supply or optional internal storage battery (courtesy Advance Bryans Instruments)

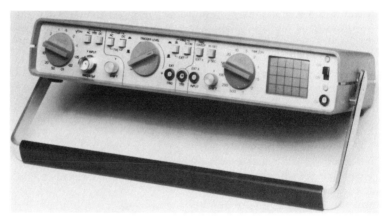

Figure 6.2 Thandar SC110 portable oscilloscope operates from internal dry batteries (courtesy Thandar Electronics Ltd)

Other oscilloscopes are capable of being powered from an external d.c. supply, e.g. a vehicle battery. *Figure 6.5* shows a typical example.

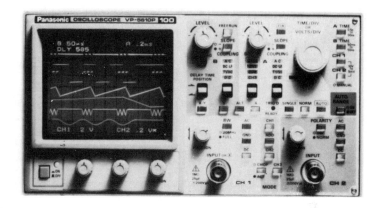

Figure 6.3 Pansonic VP–561P 100MHz oscilloscope operates from a.c. mains or external d.c. supply (courtesy Panasonic Industrial UK Ltd)

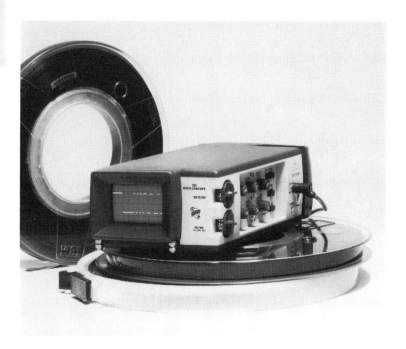

Figure 6.4 Tektronix 221 single-trace 5 MHz oscilloscope operates from internal rechargeable batteries or 90–250 V mains, a.c. or d.c. (courtesy Tektronix UK Ltd)

This instrument operates from an external d.c. supply or 100–240 V a.c. mains with any supply frequency between 50 and 400 Hz. This instrument represents a very useful compromise between a laboratory and a portable instrument, although for the latter role a 24 V d.c. supply or external battery-pack type PM8901 is required. Clearly a battery-only scope (with a small screen and limited performance) can always be made smaller and lighter than a more sophisticated mains/external d.c. powered scope.

Figure 6.5 Philips PM3050 (courtesy Pye Unicam Ltd)

Long-persistence scopes

The next important category of special-purpose scope is used for displaying low-frequency repetitive waveforms, or fast single-shot events. With the medium/short-persistence phosphors used in the majority of oscilloscopes, flicker of the trace will be noticed when its repetition rate is much lower than 50 times per second. The lower the repetition rate the worse the flicker, and at around 15 traces per second the eye ceases to see a trace at all, seeing only a moving spot of light bobbing up and down.

One solution to this problem is to use an oscilloscope fitted with a c.r.t. having a long-persistence phosphor. With this type the path traced out by the spot continues to glow for several or even many seconds afterwards. There is a wide range of different phosphors used in c.r.t.s, see Appendix 1, but one of the commonest long-persistence phosphors is type P7, with a blue 'flash' (fluorescence) and a yellowish-green 'afterglow' (phosphorescence) which fades out gradually over about eight seconds. A coloured filter suppresses the blue spot, which could otherwise be distracting as it is quite bright, leaving only the afterglow. With a long-persistence scope using this phosphor, repetition rates down to one trace per second or less can be observed; the moving spot of light is visible, but leaves its path traced out as a line of light behind it. Such an instrument also allows the observation of short single occurrences. For example, the few milliseconds of contact bounce on a switch or relay can be 'frozen' using single-shot triggering, and observed for a few seconds until the trace fades away. Several manufacturers offer a long-persistence version of one or more of their standard oscillo-scopes, see *Figure 6.6*. These long-persistence versions usually have additional extra-slow sweep speeds.

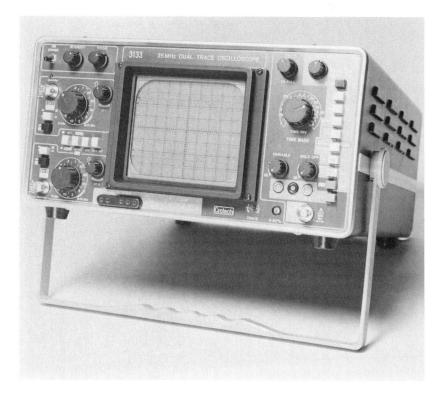

Figure 6.6   The Crotech 3133/P7 25 MHz dual channel oscilloscope with variable hold-off and long-persistence c.r.t. This provides an inexpensive means of viewing fast or infrequent single-shot events (courtesy Crotech Instruments Ltd)

There are two disadvantages to the long-persistence scope, useful though it undoubtedly is in the appropriate circumstances. The first is that the trace persists for a few seconds only and is then gradually irretrievably lost (though it can of course be photographed). The second is that if the timebase runs repetitively but without being correctly triggered, the screen rapidly fills up with a spaghetti jungle of unwanted traces that always seem to take ages to fade away.

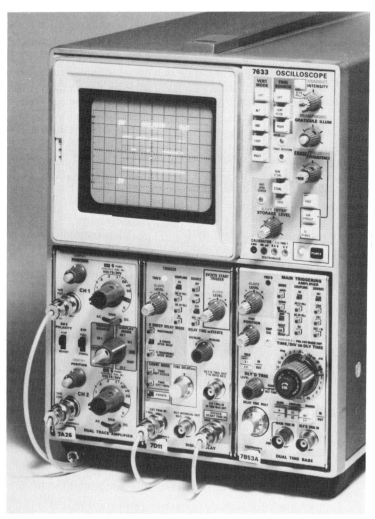

Figure 6.7 The 7633 100 MHz multimode storage oscilloscope (courtesy Tektronix UK Ltd)

Storage scopes

The ideal solution to the above problem would be an oscilloscope with variable persistence, adjustable from shortish for low-frequency repetitive waveforms to very long for single-shot traces. This is in fact exactly what a conventional storage oscilloscope does, when used in its variable-persistence mode.

Figure 6.7 shows an advanced analogue storage scope with very high writing speed: 2200 divisions/μs in the FAST VARIABLE PERSISTENCE mode with reduced scan. In addition to conventional non-storage use, the 7633 has eight storage modes: VARIABLE PERSISTENCE, BISTABLE, FAST VARIABLE PERSISTENCE and FAST BISTABLE, each at full or reduced scan. Storage time in all modes can be greater than half an hour, if the SAVE control is used when using the VARIABLE PERSISTENCE modes. Ease of use in the storage modes is ensured by the AUTOFOCUS, AUTO-ERASE and SAVE facilities.

A wide range of storage scopes is available from a variety of manufacturers. If one is content to accept a modest writing speed, a long storage time is possible; for example, the Tektronix 7613 provides storage times up to one hour at a reduced writing speed of 0.1 div/μs (compared with a maximum of 5 div/μs) with the 'save' time control in use. This instrument is a variable-persistance storage scope.

The Sony/Tektronix model 314 offers non-storage and bistable storage modes, with a storage viewing time of four hours. Shortage of space prevents a detailed discussion of the different types of storage c.r.t. available, but as the name implies, with a bistable tube the trace is either written or it isn't. In other words, if the beam current is high enough and the spot is not moving too quickly, the writing threshold is exceeded.

The advantage of this type of storage oscilloscope is that the stored trace can be retained and displayed for hours, without the gradual fading that occurs with other types of storage c.r.t. A further useful feature of some other Tektronix bistable storage scopes is the 'split screen', whereby a trace can be stored on the upper or lower half of the screen and retained while other traces can be displayed (stored or non-stored) on the other half of the screen for comparison. The price paid for the extended storage time offered by bistable storage is writing speed, which is only a few hundred divisions per microsecond at most, against up to 5000 divs/μs for fast variable persistence storage.

For viewing very low frequency repetitive or single-shot waveforms the storage scope is a far more useful and flexible tool than the long-persistence scope. However, the storage c.r.t. that it uses is far more complicated and expensive than a long-persistence c.r.t., and it also needs sophisticated control circuitry to make it work. The result is that a storage scope is a much more expensive instrument than a scope with a long-persistence c.r.t., as well as a more flexible one. As always in life, one gets what one pays for.

Digital storage scopes

With the great advances made recently in digital circuitry, especially VLSI (very large scale integration), an entirely different approach to the storage oscilloscope has become possible. The new type not only has a unique 'pre-trigger' advantage over the conventional storage scope but also does away with the delicate and expensive storage tube. *Figure 6.8* shows a product of one of the first companies in the field of digital storage oscilloscopes, the first versions of which appeared in the early seventies. Their recently introduced model 4094, *Figure 6.8*, includes twin floppy disk units. These can store numerous traces from the screen for later recall, processing (e.g. FFT) or transfer to a computer.

Digital storage oscilloscopes are now available from a number of manufacturers, and work as follows (see *Figure 6.12*). The instantaneous input voltage is sampled by a 'sample and hold' circuit and fed to an analogue-to-digital converter. Here, its value is coded in binary form, typically eight bits giving a resolution in the Y axis of better than 0.5 per cent of full screen. This digital value is stored, a new sample taken and so on, with typically 1024 samples required for one complete scan. The 1024 samples correspond to 1024 sequential positions across the screen, giving a resolution in the X direction of better than 0.1 per cent. The stored Y information is read out of the store sequentially, passed through a digital-to-analogue converter, and the resultant analogue waveform displayed on a cathode-ray tube in the usual way. However, with the digitised information representing the waveform safely in store, the storage time is effectively infinite—the information will be retained without any gradual deterioration (as occurs in a conventional storage tube) for as long as the instrument remains switched on.

For repetitive applications, in place of the sweep of an ordinary scope, the digital scope will cycle through all addresses in the X direction in turn, taking a new sample at each and replacing the previously stored Y deflection value at this address by the new value. The time taken to do this for all addresses corresponds to the sweep time for one complete trace on an ordinary scope. Clearly, if the commencement of the updating process is synchronised by a trigger circuit to the period of a repetitive Y input, a steady display will be obtained, exactly as on an ordinary scope. The big difference is that when displaying very-low-frequency waveforms, the updating of samples can be carried out at a correspondingly low rate quite independently of the X timebase, which can continue to cycle through all the addresses at a higher rate, displaying the stored information as a flicker-free trace.

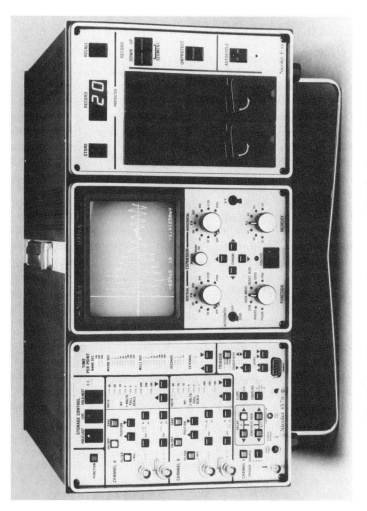

Figure 6.8 Nicolet digital storage oscilloscope model 4094 with type 4570 12 bit high resolution plug-in and 10 MS/s digitizing. Other models in the range offer sampling rates up to 50 MS/s.

The separate dots of which the trace is composed may be visible on steep edges but usually merge to form a continuous trace, while the scan (updating) can be observed if at all as a slight discontinuity moving slowly across the screen as each successive address is updated. The instrument thus behaves as a long-persistence scope with an indefinitely long persistence, with the big difference that during the course of each new scan (repetitive or single-shot) the slate is progressively wiped clean of the previous trace, which is completely replaced by the new one.

One of the advantages of the digital storage oscilloscope is that its cathode-ray tube cannot be damaged by careless misuse of the intensity and persistence controls found on a conventional storage scope. On the other hand, the maximum Y input bandwidth of a digital storage oscilloscope is limited by the speed at which the digital-to-analogue converter can work. Continuous improvements are being made in this direction; at the date of writing, digital storage scopes (or digital storage converters for use with a conventional scope as a display) are available with up to 250 MHz single shot (real time) bandwidth, though with limited six-bit (one part in 64) resolution. To achieve this bandwidth, the sample and hold circuit and the analogue-to-digital converter work at a 1000 Msamples/sec rate. Thus although digital storage scopes currently approach the bandwidth of the fastest conventional storage scopes, they can only do so with limited vertical resolution. On the other hand, instruments are available (at much lower bandwidth) with 10-bit and even 12-bit vertical resolution, corresponding to approximately 0.1 per cent and 0.025 per cent of full screen respectively.

Many instruments permit the user to store a trace at alternate addresses only, when required. This halves the horizontal resolution, but allows a trace to be retained as a reference while later traces are displayed in the remaining storage locations for comparison. On instruments with high resolution (say 10 or 12 bits vertical, with a correspondingly fine horizontal resolution) display expansion can be applied to a stored trace, enabling its fine detail to be examined at leisure after it has been captured. With the instrument shown in *Figure 6.8* up to × 256 expansion can be applied in both the vertical and horizontal directions, although at this expansion vertical resolution is low. In the expanded mode, only the selected portion of the waveform in the X direction is displayed. Similarly in the Y direction, though in this case 'off-screen' portions of the trace do not disappear but are clipped and shown as horizontal segments of trace at the top and bottom of the screen.

One of the most useful features of a digital storage scope is its pre-trigger store capability. This works as follows. If one complete presentation of the currently stored waveform is displayed for each new sample acquired (or for every nth sample), and the next trace is displayed starting from the next (or nth next) storage location, the trace will appear to run through, as though the waveform were passing before one's eyes in real time, or like a chart recorder: this is called 'roll mode'. A trigger signal, either from an external input or from the displayed waveform itself, can then *terminate* the sampling action of the A/D (analogue to digital) converter, leaving in store a segment of the input signal that occurred immediately *before* the trigger event. By adding a trigger delay, the sampling can be terminated shortly after the trigger event, leaving in store part of the signal preceding the trigger and part following, in any desired ratio. This facility is peculiar to the digital storage oscillocope, by virtue of its mode of operation, and is not available on conventional storage scopes.

Two factors are important in determining the single-shot bandwidth of a DSO (digital storage oscilloscope). Firstly (referring to *Figure 6.12)*, once a sample has been taken, the ADC requires a finite length of time to produce the digital output representing the analogue voltage presented to its input by the hold buffer amplifier. In instruments with a high digitizing rate it is common to use a 'flash' A to D converter (parallel converter) rather than the cheaper but slower successive approximation ADC. During this A to D conversion time, the sampling switch must remain open and the hold buffer amplifier must faithfully preserve the analogue voltage stored in the hold capacitor. Once conversion is complete, a new sample can be taken: this is the other factor limiting the number of samples per second and hence the single-shot bandwidth of a DSO. The control signal commands the sampling switch to close, the input buffer amplifier must charge the hold capacitor up completely to the current level of the input voltage and the sampling switch must then open, leaving the new sample stored in the hold capacitor.

Thus the maximum sampling rate is the reciprocal of the sum of the sampling time and the conversion time. The single-shot bandwidth is limited to a fraction of the sampling rate, since in general at least four or five samples per cycle of the input waveform are necessary to define it, even crudely — see *Figure 6.10.* With so few samples per cycle, the dot-joining display, trace B is a real improvement on the basic dot display of trace A. On some DSOs, one can alternatively select 'sine interpolation'. This method assumes the input signal to be a sinewave and reconstructs it accordingly from the sampled data. It offers about a 2:1 increase in the single-shot (real time) bandwidth, as it can operate with about $2\frac{1}{2}$ samples per cycle of the input, on average. Clearly, it should be used with caution, as the results could be misleading if the input waveform were not in fact sinusoidal.

The time taken for the sampling switch to change from short circuit to open circuit is called the 'aperture time' and is of crucial importance to the repetitive (equivalent time) bandwidth. Ideally, the aperture time should not be more than about 10% of the period of the input signal. As the aperture time approaches 50%, the response will be a few decibels down on the true value, depending on exactly how the sampling switch resistance changes with time as it changes from closed to open circuit.

In practice the sampling time, and in particular the aperture time, can be made much shorter than the conversion time. This permits the digitizing of a waveform whose frequency is much higher than the sampling rate, provided that the waveform is repetitive, in much the same way as the analogue sampling scope described later in this chapter. For example, the Hewlett Packard model HP54100A DSO takes up to 40 MS/s, limiting its useful single-shot bandwidth to less than 10 MHz. However, like other instruments in the HP54100 range, its aperture time is of the order of 10 ps. This enables the instrument to provide a repetitive (equivalent time) bandwidth of 1 GHz. The HP54111D, in the same range, operates at up to 1000 MS/s and this represents the current state of the art in direct digitization.

An alternative approach to high digitizing rates is to store the input analogue waveform in real time in some suitable device, and then to read it out again at some slower rate to the digitizer. One such storage device is a CCD (charge coupled device) delay line, used in the Gould series 4070 DSOs (see *Figure 6.9*). Samples of the analogue input signal are continuously fed into the CCD line at a rate set by a high speed input clock. On or following the triggering event the high speed clock is stopped, leaving the most recent analogue samples in the line. They are then clocked out of the far end of the line by the much slower output clock and applied to the ADC. At lower sampling rates, where the CCD input and output clock rates can be identical, the continuous roll-mode (chart recorder type display) is available. This technology provides a high digitizing rate at comparatively low cost. Note that it is a single-shot technique, not relying on numerous repetitions of the waveforms.

The oscilloscope illustrated in *Figure 6.20* uses an auxiliary scan converter tube as the analogue waveform store: it has a writing speed of 3000 cm/μs. The stored trace is then read out more slowly and digitized, providing an effective 25,000 MS/s digitizing rate. The Tektronix 7912AD Programmable Digitizer uses a silicon diode target array as the high speed temporary store, furnishing an effective 100,000 MS/s digitizing rate.

The latest DSOs offer many more facilities than earlier models, including both pre- and post-processing. Averaging is a useful feature when dealing with noisy signals, and works as follows. At each screen location, instead of storing the current sample in place of the previous one, the value stored would be, say, one-sixteenth of the present sample plus fifteen-sixteenths of the previous value. This provides a running average; the averaging may be made less of more pronounced by using an eighth or a thirty-second (or some other fraction) of the current sample. Averaging needs no extra storage, one storage location per screen location (for each trace) suffices. If two storage locations are allocated to each point across the screen, then a further function called enveloping can be provided. At each location, one store holds the highest value encountered to date at that point of the waveform and the other holds the lowest. The display shows the upper and lower values at each location as bright dots joined by a feinter vertical line. A stable noise-free waveform will look the same with or without enveloping, but with it, a noisy waveform will show a thickened trace; in particular, glitches will be clearly indicated.

The increasing capabilities of DSOs have led Hewlett Packard to concentrate on these entirely and abandon conventional CRT storage scopes altogether. Nevertheless there are one or two things that at present CRT storage scopes can still do better. For example, the envelope mode of the DSO described above only indicates the maximum and minimum values of the voltage observed at the given point in the waveform; it provides no information as to the distribution of values between these two extremes. The variable persistence mode of a CRT storage scope will show a noisy trace as a blurred line, brighter at the centre and trailing off in intensity above and below where the limits of variation are approached, thus providing an indication of the degree of of variability of the waveform. DSOs with pixel storage as well as wave-form storage can go some way towards this, when used in the infinite persistence mode, see *Figure 6.22*.

Aliasing

A point to bear in mind when using digital storage scopes is that they can give a misleading picture, owing to aliasing, if the input frequency is too high relative to the sampling rate; see *Figure 6.11*. They are there-fore generally specified as having a maximum bandwidth of a quarter to a fifth of the maximum sampling frequency, which is deliberately limited to allow for the finite speed of the A/D converter. The useful bandwidth at any sweep speed will be the same fraction of the sampling frequency at that sweep speed. Thus with 1024-point resolution and a 0.1 s sweep time the useful bandwidth would only be a quarter to a fifth of 1024/0.1, i.e. about 2 kHz.

Aliasing is simply the same stroboscopic effect that in a film or on TV makes the wheels of a car appear stationary or turning backwards when the vehicle is actually travelling forward. If the cine camera shutter opens and closes at the same rate as the spokes move on one space, the wheel will appear stationary. The effect is common to all sampling systems, and can be minimised by restricting the bandwidth of the signal applied to the sampling circuit to less than half the 'Nyquist rate', that is to less than half the sampling frequency. For example, eight-bit p.c.m. (pulse code modulation) telephone systems generally sample at

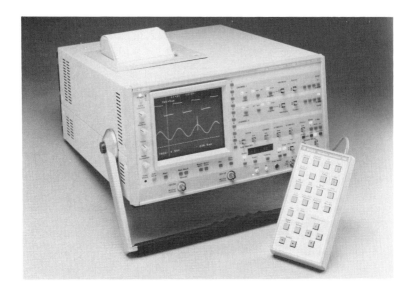

Figure 6.9 The remarkable Gould 4072 Digital Storage Oscilloscope maintains a full 8 bit (0.4%) resolution up to to its maximum sampling rate of 400 MS/s. An extensive range of post-processing facilities are available via the optional waveform processor keypad. The built in digital colour plotter provides fully annotated hardcopy output of the traces. Being fully programmable for easy integration into ATE systems, the 4072 can also transfer captured data via its computer interfaces for mass storage and specialized signal processing. The 4072 has three types of display; basic dot and pulse interpolation displays as in *Figure 6.10*, and sine interpolation display — see text (courtesy Gould Electronics Ltd)

8000 samples per second (64 kbit/s) and are fitted with anti-aliasing filters to restrict the bandwidth to less than 4 kHz. However, as we have seen, with a digital storage scope the sampling speed varies, depending on the timebase setting. Anti-aliasing filtering would therefore need its cut-off frequency switched in sympathy with the timebase speed and would add very greatly to the cost of the instrument. Thus it is usually omitted (or restricted to one frequency, and switched in or out as appropriate), and it is up to the user to be aware of the possiblity of aliasing. Either allow for it when interpreting the display, or filter the signal appropriately before applying it to a digital storage scope.

Waveform recorders

Once a digital storage scope has captured and stored a waveform it can display it indefinitely. The X and Y deflection voltages can also be made available externally for connection to an X/Y plotter, and the oscilloscope controls set to trace out the waveform quite slowly. Thus

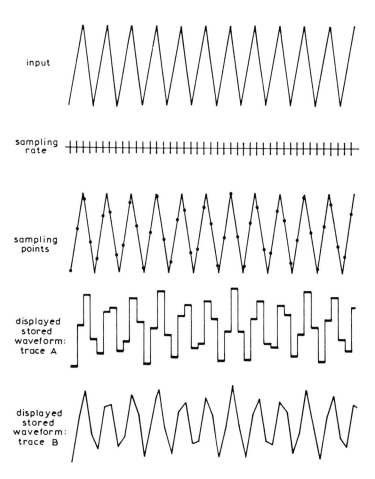

input

sampling rate

sampling points

displayed stored waveform: trace A

displayed stored waveform: trace B

Figure 6.10 Waveforms as displayed on a digital storage oscilloscope; the sampling rate is just less than four times the signal frequency. Trace A is the usual type of display: current sample held and displayed until replaced by next sample. The transitions between dots here shown as feint vertical lines would in practice be invisible. The dots themselves would be shorter than shown (with typically 1024 point horizontal resolution), except when using horizontal expansion. Trace B: same input and sample rate, but displayed using a 'dot joiner' circuit, also known as a 'pulse interpolator'

what was originally a fast waveform or transient can be reproduced in hard-copy form using an inexpensive (and hence fairly slow) X/Y plotter or chart recorder. If this recording function is all that is required, a 'digital waveform recorder' can be used; this is basically a digital storage scope without the c.r.t. display. A typical example is the

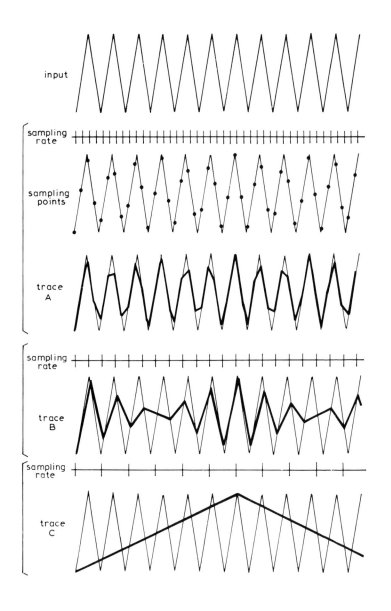

Figure 6.11 Aliasing in digital storage (and sampling) oscilloscopes. Sampling rate about four times the input frequency is just acceptable (trace A); lower sampling rate is unsatisfactory (trace B) and can be totally misleading (trace C) (courtesy Gould Instrument Division)

'Datalab Systems' model DL1080 Programmable Waveform Recorder, with both front panel and GPIB control.

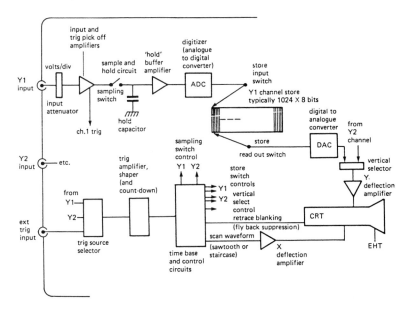

Figure 6.12 Outline block diagram of a typical DSO (Digital Storage Oscilloscope)

Sampling scopes

The system of sampling the signal at regular intervals, used in the digital storage scope, is also used in sampling scopes, and therefore the same comments as above concerning aliasing apply. The difference between a sampling scope and a digital storage scope is that, unlike the latter, the sampling scope does not apply the samples it takes to an A/D converter. The sample is stored in a 'hold' capacitor for one sampling period only, when it is replaced by the next sample. This may not occur until many cycles of the input waveform later, but with a slightly greater trigger delay for each successive sample. Thus a complete picture of the input waveform is built up over hundreds or thousands of cycles. Hence sampling scopes are only applicable to repetitive waveforms.

The main requirement for a sampling scope is a very fast analogue gate driven by a very narrow sampling pulse. Schottky diodes, avalanche transistors, pulse-forming lines and 'snap-off' diodes are components that have all been available for years, and in fact a two-channel sampling scope (the Hewlett Packard model 180) with a bandwidth of 2 GHz was available in the late 1950s. Since then the state of the art has progressed. Typical of modern sampling scopes is the standard Tektronix 7000 series mainframe with a plug-in type 7S11 and a sampling head type S4. This system provides a 25 ps risetime, corresponding to a bandwidth of 14 GHz. This sort of frequency response is obviously not easy to achieve, and involves the use of a sampling pulse of 200 ps width but with very fast trailing edge of about 20 ps.

In order to achieve the widest bandwidth and shortest risetime possible, the sampling pulse is made very narrow. Consequently the hold capacitor does not have time to charge up fully to the signal voltage; in fact it typically only reaches 10% of the full voltage. However, immediately after the sample has been taken, a positive feedback servo-loop with a gain of 0.9 bootstraps the sampling circuit up to the voltage it ought to have reached, ready for the next sample. The 90% or so of the gain provided by the servo loop is factory adjusted to allow for the sampling efficiency of its associated sampling circuit. Too little servo gain reduces the effective bandwidth and increases the risetime, while too much results in a peaked frequency response and ringing on fast edges.

One problem with sampling oscilloscopes is that of viewing the leading edge of a very fast waveform. One technique is to put the signal through a delay line following the trigger pick-off point. However, the delay line has to be of very high quality not to attenuate or otherwise distort the signal, and consequently is expensive, bulky and limited to a relatively small delay. With the Tektronix type 7T11 plug-in, this difficulty is ingeniously avoided by a technique known as random sampling. Here, a special automatic circuit predicts the arrival of the next leading edge on the basis of the constantly monitored signal repetition rate, and starts the sampling process sufficiently ahead of the predicted time to catch all of the leading edge.

Time domain reflectometers

Closely allied to the sampling scope is the time domain reflectometer (TDR). This is a special-purpose sampling scope with a built-in generator that produces a repetitive step waveform with a very fast rise time, less than 35 ps in the case of the system illustrated in *Figure 6.13*. This consists of a Tektronix 7000 series mainframe with a 7S12 TDR plug-in. The fast step waveform is produced by an S52 pluse-generator head, and is applied to the piece of cable or whatever is to be tested. If the cable has the same characteristic impedance as the TDR generator throughout its length (50 Ω) and is correctly terminated at the far end, there will be no reflected step returning to the TDR. If, however, the

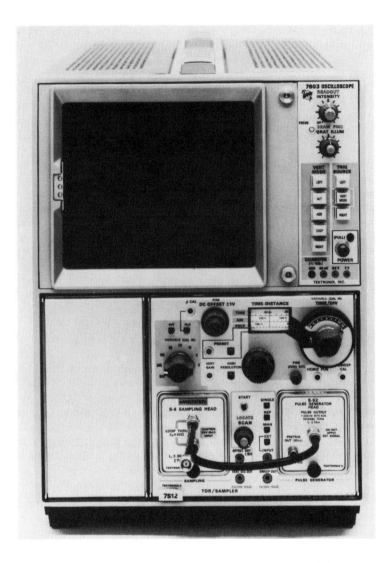

Figure 6.13　Time domain reflectometer consisting of Tektronix 7603 mainframe with 7S12 plug-in containing S52 pulse generator and S6 sampling heads (courtesy Tektronix UK Ltd)

cable is damaged or incorrectly terminated a reflection will originate from the point where the mismatch occurs and will travel back towards the TDR. This reflected signal is sampled by the S6 sampling head, both heads being accommodated within the 7S12—plug-ins within a plug-in.

Overall rise time of the displayed return, including the effect of the S6 head, is 45 ps, giving a maximum resolution of about 0.5 cm.

The purpose of a TDR is to measure the magnitude and position of a discontinuity (or even several discontinuities, but the reduced magnitude accuracy) in a coaxial system, say a cable or a step attenuator. In the horizontal direction the display measures time (as on a normal scope) and also the distance to the discontinuity, as this is a direct function of time, given the velocity of propagation of signals in air, polythene, or whatever the insulation of the cable is made from. In the vertical direction the display reads ρ (reflection coefficient, represented by the Greek letter 'rho') from 0 at the middle of the screen, representing a perfect match, to 1 at the top and bottom, representing total reflection of the incident step waveform, i.e. an open or short circuit respectively. Thus if the horizontal scale were set to 2 m and 1 m of 50 Ω coaxial cable were connected, the cable being terminated in a 50 Ω load, a straight line would be obtained right across the screen; but if the load were changed to 75 Ω a step would be obtained halfway across the screen, the ρ reading changing abruptly from zero to 0.2. This corresponds to a voltage standing wave ratio (v.s.w.r.) of 1.5, since v.s.w.r. $= (1 + \rho)/(1 - \rho)$.

Various sweep speeds provide for a wide range of maximum cable lengths up to 120 metres of air-insulated line. For the measurement of the magnitude of very small discontinuities, the Y axis can be expanded up to ± 2 mρ per division (i.e. 0.002 reflection coefficient per division), permitting a reflection coefficient as low as 0.001 to be measured. A reflection coefficient of 0.001 corresponds to a v.s.w.r. of 1.002, a very small mismatch indeed. Owing to the effect of multiple reflections, measurement accuracy of ρ is degraded where several discontinuities are present on a cable run. The distance indicated to the first (nearest) discontinuity is accurate though, and when this is rectified the next discontinuity can be accurately resolved and so on.

The 45 ps resolution of the system corresponds to 0.66 cm in air or about 0.45 cm in a practical transmission line, allowing testing not only of cables but also of the path through various components, e.g. step attentuators, coaxial switches, coaxial relays, etc.

Spectrum and logic analysers

So far the purpose of all the instruments described in this chapter has been to display voltage waveforms as a function of time, although in the case of the TDR this is admittedly only as a means to an end. We now come to a class of measuring instruments that use a cathode-ray tube to display the measured results, but are not strictly oscilloscopes at all in the above sense. However, they are included here for two reasons. First, they are very important tools and hence interesting in their own right; second, they frequently take the form of one or more plug-in units fitting in an oscilloscope mainframe.

Spectrum analysers

Spectrum analysers display voltage in the vertical direction, but in the horizontal direction the baseline is not time but frequency. They are particularly useful for analysing very complex signals, especially where several components of unrelated frequencies are present simultaneously. Each frequency component appears as a vertical line at the appropriate position along the baseline. Each vertical line reaches up from the baseline to a height dependent on the r.m.s. amplitude of that frequency. The individual components are thus sorted out, whereas a conventional oscilloscope cannot produce a usable steady display from an input containing several components of unrelated frequencies.

Thus a spectrum analyser is said to work in the frequency domain, unlike an oscilloscope, which works in the time domain. This means simply that the horizontal or X axis of the former is calibrated in frequency per division, whereas that of the latter is calibrated in time per division. In the vertical or Y direction the spectrum analyser displays the amplitude of each of the individual frequency components, measured by a selective receiver swept across the band of interest. The amplitudes of any signals it encounters are measured by a detector circuit, much as in any ordinary superhet radio receiver. The modern spectrum analyser is in fact just an improved version of the older panoramic receiver, which did much the same job but was usually not very accurately calibrated.

As the signals applied to the spectrum analyser may vary widely in amplitude it is usual for the vertical axis to display these amplitudes logarithmically. Thus typically one vertical division equals 10 db (10 decibels means a power ratio of 10:1 or a voltage ratio of 3.16:1). If the top line on the screen represents 0 dBm (i.e. an input power of 1 milliwatt or 225 mV into a 50 Ω load) and two signals are present one reaching to the top of the screen and one only to three divisions down, the smaller signal is said to be '30 dB down' on the larger. Its power is thus one microwatt or 225 mV divided by $(3.16)^3$ or 7.1 mV. In the 10 dB per division mode a good spectrum analyser can resolve signals down to 70 dB or more below full screen, even in the presence of other full-screen signals. Input and intermediate-frequency attenuators usually enable the full-screen level to be set in 10 dB steps from +20 dBm (100 mW) down to −40 dBm or lower. Similarly, in the horizontal direction, the frequency corresponding to the vertical centre line can be set anywhere within the range covered by the instrument. The 'dispersion', i.e. the frequency span per horizontal division, can be set within wide limits to cover as much or as little bandwidth either side of the centre frequency as required.

A logarithmic display with 10 dB/div is useful for coping with signals of widely differing amplitudes, but for some purposes extra vertical resolution is desirable. An example would be when checking the upper and lower sidebands of the output of a frequency-modulated signal generator for equality, to confirm the absence of incidental amplitude modulation. For such applications, spectrum analysers generally provide a 'linear' mode, where the vertical height of a displayed signal is directly proportional to its amplitude, and sometimes also a 2 dB/div mode. In both cases, the extra resolution is bought at the cost of reduced on-screen dynamic range.

Figure 6.14 shows a Hewlett Packard microwave spectrum analyser plug-in type 8559A, fitted into a 182T-type mainframe. This instrument covers 10 MHz to 21 GHz, with dispersion adjustable from 1 kHz/div to 1 GHz/div. Resolution bandwidths available are 1 kHz to 3 MHz in a 1, 3, 10 sequence. Most measurements can be quickly made using only three controls without sacrificing accuracy, owing to coupling of the analyser's controls. For example, as span width is reduced for more detailed analysis, the resolution bandwidth, video filter and sweep time automatically change to the optimum values for a calibrated display. This prevents erroneous readings, which can occur on instruments without interlocked controls when these controls are incorrectly set. A moment's reflection will make it clear that if a spectrum analyser is set to a narrow i.f. bandwidth and made to sweep a wide frequency range at too high a repetition rate, any signals encountered will not remain within the i.f. bandwidth long enough to produce an accurate display of their full amplitude. Most spectrum analysers have warning lights to alert the user, under these conditions, that he must either reduce the sweep width (span) or its repetition rate, or use a wider i.f. and/or 'video' (i.e. detector smoothing) bandwidth. Increasingly, modern spectrum analysers such as the HP8559A have interlocked controls to prevent such mis-setting.

The Hewlett Packard range of spectrum analysers includes types covering the audio-frequency range (type 3580A, 0–50 kHz with digital storage display and adaptive sweep) to the microwave range (type 8565A, covering 10 MHz to 22 GHz, or to 40 GHz with external mixer).

Many other companies, mostly American or Japanese, also produce spectrum analysers. One notable UK-manufactured spectrum analyser is the Marconi Instruments type 2382, covering 100 Hz to 400 MHz with digital storage display and interlocked controls, shown in *Figure 6.15*. This has a built-in sweep generator whose frequency is identical to the spectrum-analyser section's receive frequency. In conjunction with a split-display feature this forms a powerful tool for the Test Department. For example, with the levelled output of the sweeper

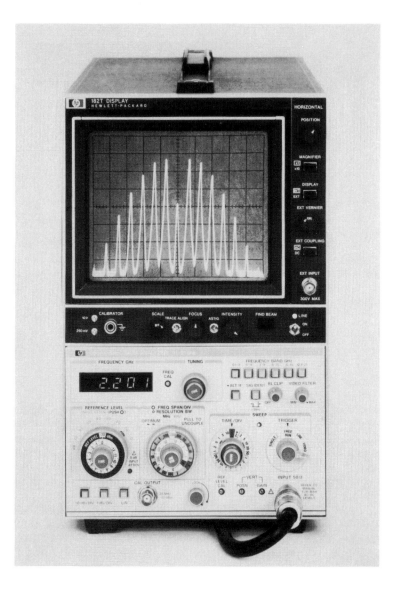

Figure 6.14 Hewlett Packard 8559A spectrum analyser plug-in (in 182T main-frame) covers 10 MHz to 21 GHz (courtesy Hewlett Packard Ltd)

connected to the spectrum analyser via a correctly aligned filter, the latter's frequency response can be stored using one half of the display. Unaligned filters straight from production can then be substituted, and

Figure 6.15 Marconi Instruments 2382 spectrum analyser with digital storage display covers 100Hz to 400MHz. It features digital memory, resolution bandwidth down to 3Hz and video (post detection) bandwidths down to 1Hz, with video averaging up to 128 traces and 100dB display range (courtesy Marconi Instruments Ltd)

their response compared using the other half of the display with the reference trace. It is then a simple matter to set up an unaligned filter by adjusting it so that its response is superimposed exactly on that of the reference trace.

Logic state analysers

In the earlier days of the development of digital systems, circuit designers and field engineers alike had to make do with a dual-trace or four-trace oscilloscope, but with the explosion of electronics in the minicomputer and microprocessor era, something more suitable was needed. Eight-trace scopes with the ability to recognize an eight-bit trigger word soon appeared. With the addition of further facilities, these have evolved into the LSA. Some of the latest and most advanced logic state analysers, such as the Philips PM3570 (with up to 115 channels and up to 400 MHz acquisition rate), have more in common with a visual display unit (VDU) than with a conventional real-time oscilloscope, in that they use a large-screen cathode-ray tube with magnetic deflection. This is the display technology used in TV sets and VDUs, and is quite different

Figure 6.16 Philips PM 3540 'Logic Scope': 25 MHz dual-trace real-time oscilloscope combined with a logic analyser giving binary, octal or hex readout (courtesy Pye Unicam Ltd)

from the high-speed electrostatic deflection type of cathode-ray tube used in conventional oscilloscopes (see Chapter 7). Consideration here is therefore mainly of the type of LSA derived from the conventional oscilloscope.

An interesting example is the Philips model PM3540 'logic oscilloscope', a combined 25 MHz dual-trace oscilloscope and 16-channel 10 MHz logic analyser; see *Figure 6.16*. The instrument acts as oscilloscope only, as logic analyser only, or as both together. The logic analyser provides a choice of binary, octal or hexadecimal state display on the oscilloscope screen. Features include 'compare mode' operation, 64-word memory with 16-word display, clock qualifiers, and versatile trigger facilities including delay. Trigger extension is possible with trigger probes. Comprehensive trigger facilities are provided-internally, externally or manually. A trigger word can be preset in the desired display format. This can be done either by using the display and cursor controls or by selecting the desired condition from the data stream itself to allow fast paging through the data stream.

Sampling is synchronous up to 10 MHz. Threshold is variable with a fixed TTL level position. Start can be either manual or automatic with the instrument acquiring and displaying data continuously. The trigger can be delayed for up to 9999 pulses.

Main memory is 64 × 16 bits, with a display on the oscilloscope tube of 16 × 16 bits. Cursor and column-blank facilities are provided. The compare mode uses two separate 64-word memories. A state table can be stored in one memory for comparison with new collected data. Inequalities are indicated in the heading, and show up as intensified digits.

There are 16 input channels for the logic analyser display. Input is provided by three sets of eight miniature probes, which also supply an external clock input and three clock qualifier inputs.

The oscilloscope section provides a high 2 mV sensitivity over the full 25 MHz bandwidth. It has two vertical channel inputs completely separate from those of the logic analyser. A wide choice of triggering facilities is provided, including peak-to-peak automatic and d.c. coupling. Trigger sources can be either vertical input, composite, external or the logic analyser.

A wide variety of logic state analysers is available from a number of companies. Hewlett Packard claim to have introduced the first LSA in 1973. *Figure 6.18* shows the 1631D, from their current range. This provides up to 43 state channels operating at up to 25 MHz and can acquire up to 1024 words using various triggers and qualifiers to initiate capture. The captured activity on address, data, status and control lines

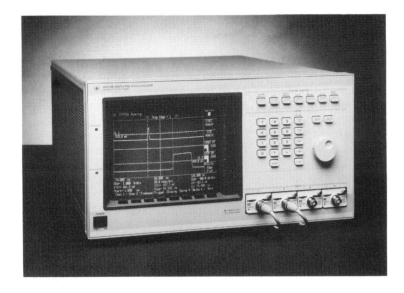

Figure 6.17 Hewlett-Packard Company's first colour-display digitizing oscilloscope, the HP 54110D, provides 1 GHz bandwidth and a feature set designed for design and test engineers working with high-speed logic and high-speed data communications (courtesy Hewlett Packard Ltd)

may be displayed in binary, octal, decimal, hexadecimal, ASCII or mnemonics. Up to 16 channels operating at up to 100 MHz may be dedicated to timing capture and the usual reconstructed digital waveforms displayed; glitches are captured and displayed as stereotypes, as at the x cursor in *Figure 6.21*. In addition there are two analog input channels, the 200 Msamples/sec sampling rate providing a 50 MHz bandwidth. Any two of the modes may be used interactively. Thus full state analyser indexing can arm analog waveform acquisition or vice versa. Similarly, the state analyser and timing modes can be used together, as can the timing and analog modes, *Figure 6.21*. In both analog and timing modes, user definable trigger conditions can be set to terminate data acquisition and 'post-processing' time interval measurements made on the stored waveforms with the aid of two timing cursors.

In addition to stand-alone logic analysers such as the HP1631D, logic analyser plug-ins are available for mainframe oscilloscopes. The Tektronix 7D01 logic analyser is a double-width plug-in that fits in any three- or four-compartment 7000 series oscilloscope mainframe; see *Figure 6.19*. In a four-compartment mainframe it may be used in conjunction with

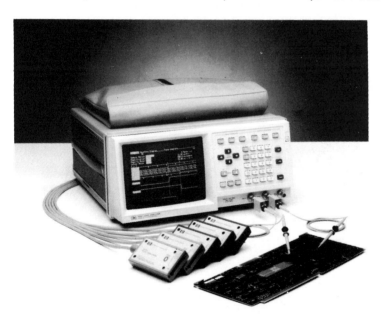

Figure 6.18   The Hewlett Packard 1630 series logic analysers offer timing analysis – with glitch capture – for the hardware man plus state and performance analysis for the software man. To this, the 1631D (shown above) adds dual channel digital storage oscilloscope functions at up to 200 Msamples/sec, giving 50 MHz bandwidth (courtesy Hewlett Packard Ltd)

X and Y plug-ins to provide a logic analysis capability combined with standard oscilloscope functions in a single instrument. Data is captured by probes with TTL or adjustable thresholds and displayed in one of three formats: 16 channels by 254 bits, 8 x 508 or 4 x 1016. Asynchronous data sampling rates up to 100 MHz (four-channel only) may

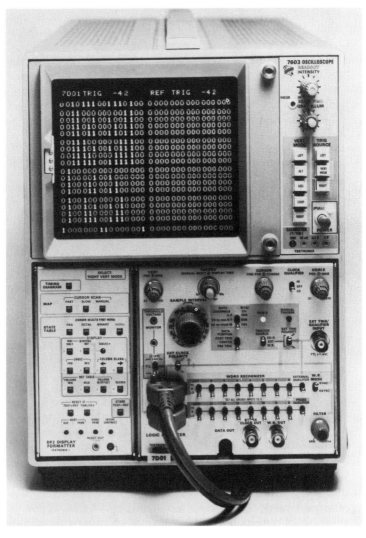

Figure 6.19 Tektronix 7D01 logic analyser and DF2 display formatter in type 7603 mainframe (courtesy Tektronix UK Ltd)

be selected, providing timing resolution to 15 ns. Data can also be synchronously (externally) clocked to 50 MHz. The built-in word recogniser may also be used to trigger the analogue portion of the oscilloscope. Probe- or external-qualifier functions select the logic state of external signals that enables word recognition. The trigger and cursor both appear as intensified spots on the display, which is in the form of a timing diagram. The number of sample intervals between the trigger and the word selected by the adjustable cursor appears as an alphanumeric readout. The word selected by the cursor is displayed as ones and zeros in three-bit or four-bit groups.

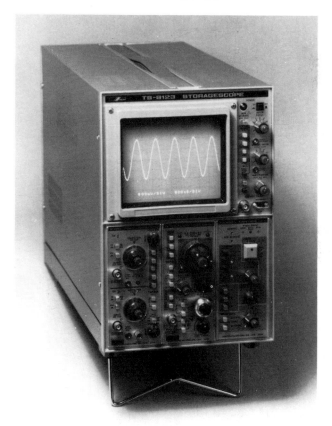

Figure 6.20 The remarkable Iwatsu TS8123 provides real-time and storage operation at writing speeds up to 3000cm/μs using a scan converter tube. The stored trace may be digitized (providing effectively a 25,000 Msample/s digitizing rate) and read out to a pen recorder or via the GP–IB interface for digital processing (courtesy STC Instrument Services)

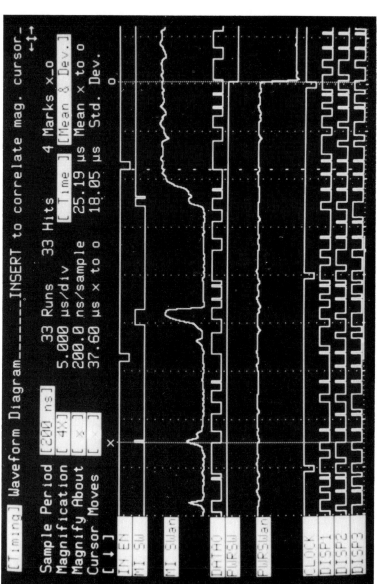

Figure 6.21 The HP 1631A/D can acquire data simultaneously in the state, timing, and analog domains. This allows events of interest in one domain to be time-correlated to events in the remaining two. In this photograph the third waveform (MI SWan) is the analog representation of the timing waveform immediately above it (MI SW). (courtesy Hewlett Packard Ltd)

When the DF2 Display Formatter (a single-width plug-in) is used in conjunction with the 7D01, no less than seven display formats are available. The timing display provides the usual simulated 16-trace oscilloscope display. The binary display mode is used to display stored data in a state table of ones and zeros. The hexadecimal display mode displays the stored data in hexadecimal (0 to F) format, and the octal mode displays it in octal (0 to 7) format. A map mode is provided and this displays up to 256 consecutive 16-bit words at a time. These can be scanned automatically or manually, with a '+' symbol used as a cursor, in the same order they were entered in memory. The word at the cursor location is identified in either hexadecimal, octal or binary at the bottom of the display. The GPIB mode monitors and displays activity on the GPIB data bus, transfer bus (handshake lines), and management bus (control lines). Disassembled instructions are displayed on the c.r.t. in IEEE 488 message mnemonics familiar to the GPIB user. The ASCII mode displays stored data in ASCII format. The ASCII characters are further decoded into hex ($ precedes decode), octal (@ precedes decode), or binary. The decode of the ASCII character is determined by the state-table mode previously selected.

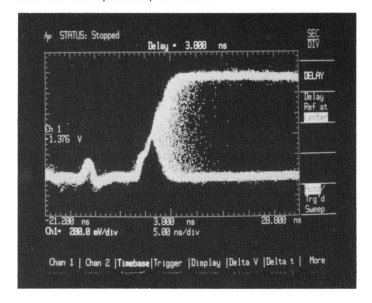

Figure 6.22   The HP54000 series DSOs use a full colour graphics type display with pixel storage. This makes it possible to store and display all voltage levels occurring at each point along a waveform over many repetitions. This is called the 'infinite persistence' mode and the illustration above shows its use to display intermittent partial transitions and double transitions at the output of a D flip-flop (courtesy Hewlett Packard Ltd)

97

7

How oscilloscopes work (1): the c.r.t.

Many logic analysers and some DSOs use magnetically deflected c.r.t.s either monochrome (as in the Thandar TA3000 High Performance Modular Logic Analyser) or colour (Hewlett Packard HP54111D DSO). This is the type of display technology used in TV sets. The operation of TV type tubes is well covered in other volumes in the Heinemann Newnes series, to which the reader is referred for further information. In c.r.t. storage oscilloscopes, the cathode ray tube is basically similar to the electrostatically deflected type of tube described in this chapter, but with the addition of one or more storage meshes: storage tubes are briefly described in Chapter 6.

This chapter deals solely with the high performance c.r.t.s using electrostatic deflection, used in non-storage oscilloscopes. Such an oscilloscope may also include a digital storage capability, as in *Figure 1.5*, and the same c.r.t. is then used for both the conventional real time display and for the storage mode display.

The cathode-ray tube is the main component of an oscilloscope. A cathode-ray tube consists basically of an electrode assembly mounted in an evacuated glass vessel (*Figure 7.1*). The electrodes perform the following functions:

• A triode assembly generates the electron beam, originally called the 'cathode ray'. It consists of a cathode K heated by a filament F, a control grid G and the first beam-acceleration electrode (1).
• An electrode (2) focuses the beam.
• The beam is then further accelerated before reaching the deflection plates.

- The vertical deflection plates change the direction of the beam in proportion to the potential difference between them. When this is zero, i.e. the two plates are at the same potential, the beam passes through undeflected. The vertical deflection plates are so called because they can deflect the beam in the vertical direction, so that it hits the screen at a higher or a lower point; they are actually mounted horizontally above and below the beam, as shown in *Figure 7.1*. Similarly the horizontal deflection plates permit the beam to be deflected to left or to right.
- The deflected beam then hits the fluorescent coating on the inner surface of the glass screen of the c.r.t. The coating consists of a thin layer of 'phosphor', a preparation of very fine crystals of metallic salts

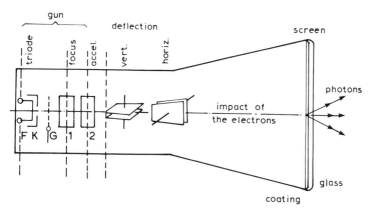

Figure 7.1   Basic oscilloscope (electrostatic) cathode-ray tube (courtesy Enertec Instrumentation Ltd)

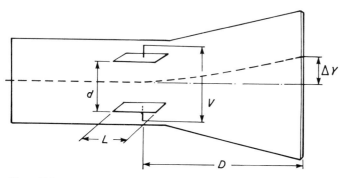

Figure 7.2   Y-deflection sensitivity: see text (courtesy Enertec Instrumentation Ltd)

99

deposited on the glass. Further details of phosphors are given in Appendix 1. The 'spot' or point of impact of the beam glows, emitting light in all directions including forwards. Modern c.r.t.s are aluminised, i.e. a thin layer of aluminium is evaporated on to the rear of the coated screen. The electrons pass through this with little retardation, causing the phosphor to glow as before, but now the light emitted rearwards is reflected forwards, almost doubling the useful light output.

The potential at the focus electrode is adjusted to obtain a very small round spot on the end of the tube. Unfortunately, if no other control were provided, it would often be found that the focus control setting for minimum spot width was different from that for minimum spot height. This is avoided by providing an astigmatism control. In the case of a simple cathode-ray tube this consists of a potentiometer that adjusts the voltage on the final anode and screen relative to the deflection plate voltages. Alternate adjustments of the focus and astigmatism controls then permit the smallest possible spot size to be achieved. With more complicated tubes using a high 'post-deflection acceleration ratio' another electrode is often needed. This is a 'geometry' electrode and is connected to another preset potentiometer, which is adjusted for minimum 'pincushion' or 'barrel' distortion of the display.

When an electron beam passes between two horizontal plates that have a potential difference of V volts between them (*Figure 7.2*) it is deflected vertically by an amount:

$$\Delta Y = \frac{KVLD}{2V_a d}$$

where L = length of the plates

D = distance between the plates and the point on the axis where the deflection is measured

d = distance between the plates

V_a = acceleration voltage applied to the beam at the level of the plates

K = a constant relating the charge of an electron to its mass.

The Y deflection sensitivity of a c.r.t. is defined by $\Delta Y/V$ and is expressed in cm/V. However, in practice the inverse relationship is normally used: $V/\Delta Y$, in V/cm, i.e. the differential deflection-plate voltage necessary to achieve a spot deflection of 1 cm.

Brilliance or intensity modulation (also called Z modulation) is obtained by the action of a potential applied to the cathode or grid that controls the intensity of the beam. Generally, a change of 5 V will produce a noticeable change of brightness, while a swing of about 50 V will extinguish a maximum-intensity trace. The beam is normally extinguished during 'flyback' or 'retrace'; see Chapter 8. This may alternatively be achieved in some c.r.t.s by means of an auxiliary 'blanking' electrode, which can deflect the beam so that it no longer passes through the deflection plates and hence does not reach the screen.

Tube sensitivity

The deflection plates of a c.r.t. are connected to amplifiers, which can be of relatively simple design when the required output amplitude is low; it is therefore desirable for the tube sensitivity to be as high as possible. To enable the amplifier to have a wide bandwidth, the capacity between the plates must be kept low, so they must be small and well

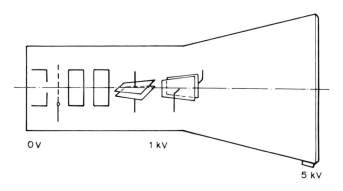

Figure 7.3 Single-stage post-deflection acceleration (courtesy Enertec Instrumentation Ltd)

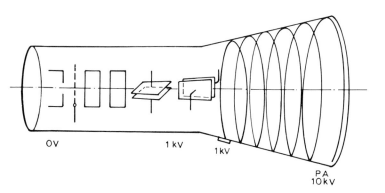

Figure 7.4 Spiral p.d.a. (courtesy Enertec Instrumentation Ltd)

separated. On the other hand, in order to obtain a suitably clear trace of a signal with low repetition frequency (or single-shot) the energy of the beam must be high. But the ideal tube must be:

- Short (not cumbersome): D small
- Bright (high acceleration voltage): V_a large
- And with low deflection-plate capacity: L small, d large

This gives a tube with very low sensitivity, considering the formula:

$$\text{Sensitivity} = \frac{\Delta Y}{V} = \frac{KLD}{2V_a d}$$

The requirements for high sensitivity contradict the terms of the equation. Practical cathode-ray tubes are therefore the result of a compromise. However, techniques have been developed to improve a selected parameter without prejudice to the others. Post-deflection acceleration (p.d.a.) is one of these; see *Figure 7.3*. To improve the trace brightness while retaining good sensitivity, it is arranged that the beam passes through the deflection system in a low energy condition (relatively low initial acceleration); post-deflection acceleration is then applied to the electrons. This is achieved by applying a voltage of several kilovolts to the screen of the c.r.t.

Spiral p.d.a., *Figure 7.4*, is a development of the basic p.d.a. technique, and consists of the application of the p.d.a. voltage to a resistive spiral (500 MΩ) deposited on the inner tube surface between the screen and the deflection system. The uniformity of the electric field is improved, which reduces distortion. In addition the effect of the p.d.a. field between the deflection plates is weaker, so the loss in sensitivity caused by this field is reduced.

The use of a field grid—*Figure 7.5(a)*— avoids any reduction in sensitivity caused by the effect of the post-deflection acceleration field. A screen is interposed between the deflection system and the p.d.a.; this makes the tube sensitivity independent of the p.d.a., a significant benefit. The screen must, of course, be transparent to the electrons and is formed from a very fine metallic grid. With this system we reach the domain of modern cathode-ray tubes.

The next development is the electrostatic expansion lens—*Figure 7.5(b)*. By modifying the shape of the field grid (e.g. a convex grid) it is possible to create, with respect to the other electrodes, an electric field that acts on the electron beam in the same way as a lens acts on a light beam. It is therefore possible to increase the beam deflection angle, for example by a factor of two, which improves the sensitivity by the same amount.

The field can also be formed by quadripolar lenses. So, for example, if the sensitivity of a spiral tube is 30 V/cm in the X axis and 10 V/cm

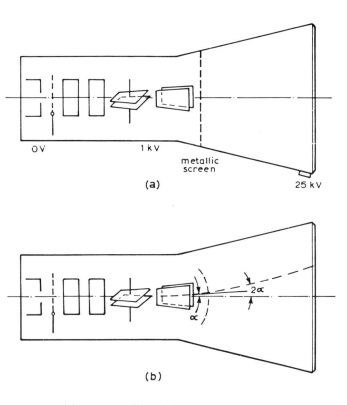

Figure 7.5 (a) Mesh p.d.a. (b) As (a) but combined with expansion lens (courtesy Enertec Instrumentation Ltd)

in the Y axis, then the sensitivity of a lens-fitted tube, for the same trace brightness, may be 8 V/cm in X and 2 V/cm in Y or even better.

To improve the sensitivity by modifying the deflection system it is necessary to do one of two things:

● Reduce the distance between the plates, increasing the capacity between them; in addition it must be possible to deflect the beam without it striking them.

● Lengthen the plates, again increasing the capacity; however, the transit time involved limits the application of this idea.

The transit time is the time taken for an electron to pass through the deflection system: $t_0 = L/$electron speed. Suppose that a sinusoidal voltage of period t_0 is applied to the deflection plates. An electron leaving the plates will be in the same position as one entering the system, because the instantaneous value of the voltage applied to the

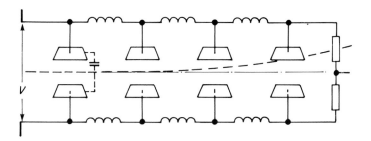

Figure 7.6 Delay-line Y-deflection plates (courtesy Enertec Instrumentation Ltd)

plates will be the same (one period between the input and the output) and there will be no deflection. To enable the beam to be deflected so as to trace the outline of the applied signal, the length of the plates must be small compared with the distance the electrons travel during the period of one cycle of the signal. So for high-frequency work the plates must be short, which again reduces the sensitivity.

This problem can be circumvented by the use of sectional plates (*Figure 7.6*). To improve the sensitivity several plates are placed in series, connected by a delay line. As the propagation velocity of the line is made equal to the speed of the electrons in the beam, the deviation accumulates successively. On the other hand the parasitic capacitance of the plates is incorporated in the delay line, which must be terminated in its characteristic impedance. The design of the line is entirely determined by its stray capacitance and the propagation time. This brings us to delay-line deflection plates (*Figure 7.7*). Here, the dimensions of the plates have been reduced and their number increased. Two flattened helices are used, each turn acting as a deflection plate. The helix is constructed in such a way that its propagation velocity corresponds to the speed of the electron beam. These deflection systems, together with field grids or quadripolar lenses (or both), permit the construction of very high-performance tubes.

Other tube characteristics

To be suitable for use at high frequencies a c.r.t. must, as already discussed, have a highly developed deflection system. But this alone is not sufficient when it is required to observe and photograph fast pulses with low repetition rates or single-shot phenomena. The brilliance of the display must also be adequate. This is why 'writing speed' is an important feature in these conditions. Writing speed is defined as the maximum speed at which a spot, passing once across the tube face, can

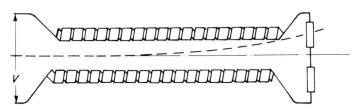

Figure 7.7 Travelling-wave Y-deflection plates (courtesy Enertec Instrumentation Ltd)

be photographed under specified conditions (camera, aperture, image/object, film sensitivity).

On the occasions when it is necessary to compare several fast, single-shot phenomena occurring simultaneously, the only solution is to use an oscilloscope equipped with a c.r.t. with several beams. There are a number of different types available:

● Multi-gun tubes. *Figure 7.8(a)* shows a c.r.t. with several cathode-ray assemblies mounted in a single tube. Part *(b)* shows a tube where each gun or triode assembly has its own vertical deflection system but shares common horizontal deflection plates. All phenomena are displayed with the same sweep speed.

● Multi-beam tubes. There is a single electron gun for the different deflection systems, typically two. The beam is shared between each deflection system by means of a splitter plate, see frontispiece. This type of tube is more economical because there is a single gun assembly. However, there is reaction between the two systems, and the brilliance of the displays cannot be adjusted separately.

Figure 7.9 shows the construction of the electrode assembly of a mesh p.d.a. cathode-ray tube. The deflection plates are within the cylindrical shield and the mesh covers the square opening at the end. The wires of which the mesh is woven are so fine that it is invisible; this also ensures that it is transparent to the beam of electrons. *Figure 7.10* shows a high-performance oscilloscope c.r.t. with side connectors to the deflection plates for minimum capacitance, spiral p.d.a., internal graticule, bonded implosion guard and light guide for graticule illumination.

All the measures to maximise the bandwidth of a c.r.t. mentioned previously—p.d.a, delay-line deflection plates, scan expansion lenses—have been put together in the cathode-ray tube used in the Tektronix type 7104 oscilloscope. This instrument boasts a 1 GHz real-time bandwidth, this limit being set by the Y amplifier rather than the c.r.t. itself. The latter could display signals up to 2.5 GHz, were it possible to design suitable wideband drive circuitry. Also, notwithstanding the conflict, explained earlier, between tube design parameters for opti-

105

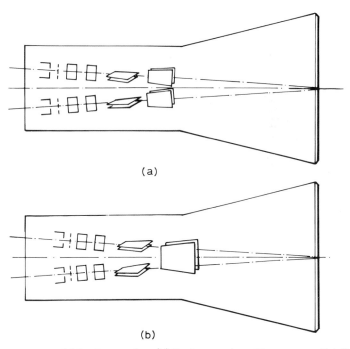

(a)

(b)

Figure 7.8 (a) Dual-gun tube. (b) Dual-gun tube with common X-deflection plates (courtesy Enertec Instrumentation Ltd)

mum bandwidth and maximum writing speed, this tube achieves the remarkable writing speed of 20,000 cm/μs, using ASA 3000 film without fogging. (In fact, single-shot events at that speed can also be seen comfortably with the naked eye.) The secret is revealed in *Figure 7.11*, which shows that in addition to the measures already mentioned, the c.r.t. incorporates a microchannel electron multiplier plate. This consists of thousands of short parallel tubes, each coated internally with a high-resistance film. Each individual tube acts as an electron multiplier by virtue of secondary emission, resulting in 10,000 electrons hitting the phosphor for each electron in the beam. Owing to the small spacing between the microchannel plate output side and the aluminised phosphor, together with the high potential difference between them, there is negligible spreading of the output of each microchannel tube, maintaining a small sharp spot size.

Figure 7.9 Electrode assembly of 'Brimar' mesh p.d.a. c.r.t. type D13-51GH (courtesy Thorn Brimar Ltd)

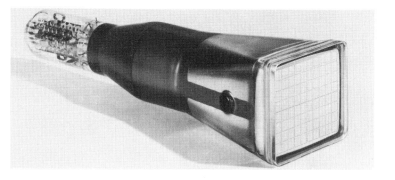

Figure 7.10 'Brimar' spiral p.d.a. c.r.t. type D14–210GH/82 with internal graticule (courtesy Thorn Brimar Ltd)

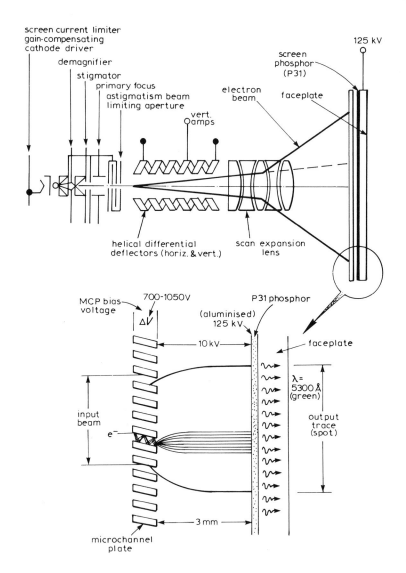

Figure 7.11 The MCP (microchannel plate) cathode ray tube used in the Tektronix oscillocope type 7104. MCP c.r.t.s are also used in the model 11302 mainframe (*Figure 3.2*) and in the 2467. This model, similar in most other respects to the 2465A (*Figure 3.1*), for the first time, enables an isolated glitch only nanoseconds wide to be seen on a portable oscilloscope (courtesy Tektronix UK Ltd)

8

How oscilloscopes work
(2): circuitry

Figure 8.1 shows the block diagram of a typical dual-trace, high-performance oscilloscope. Two identical input channels A and B are switched alternately to a common amplifier, which drives a delay line.

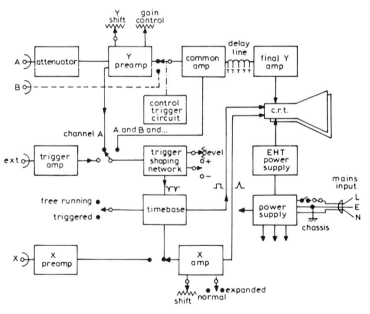

Figure 8.1 Block diagram of dual-trace mains-operated oscilloscope (courtesy Enertec Instrumentation Ltd)

This is shown diagrammatically as composed of discrete inductors and capacitors, although in a modern instrument it would usually consist of a length of delay cable. This is similar to co-axial cable, except that it has a centre conductor wound in the form of a spiral and hence provides much greater delay per unit length. As the drive to the trigger circuit is picked off before the delay line, the delay introduced by the latter permits the whole of the leading edge from which the scan was triggered to be observed. This assumes of course that the rise time of the leading edge and the 'wake-up time' of the trigger circuit are together less than the delay introduced by the delay line, which is generally tens of nanoseconds.

The final Y amplifier produces the push-pull voltages that drive the Y plates, and in a higher-performance instrument the peak-to-peak output swings required might be little more than a few tens of volts, especially if using a tube with a high p.d.a. ratio and a scan-expansion lens. The X amplifier has to provide several times as much voltage swing as the Y amplifier, as the X-plate sensitivity is less than that of the Y plates. Fortunately, a substantially smaller bandwidth suffices for the X amplifier, easing the circuit design problems; the c.r.t. designer takes advantage of this to maximise the Y-plate sensitivity at the expense of the X-plate sensitivity.

The X deflection amplifier is driven with a sawtooth waveform produced by a 'sweep' or 'timebase' generator, which itself is triggered by a pulse from the trigger circuit. The trigger circuit produces a pulse each time the Y input voltage crosses a given threshold voltage, which is usually adjustable by the front-panel trigger level control. Thus the sweep always starts at the same point on the waveform, the sweep generator thereafter being insensitive to further trigger pulses until it has completed both the trace and the following (blanked) 'retrace' or 'flyback'.

Circuit elements

It is probably true to say that, at the time of writing, the majority of oscilloscope designs make use mainly of discrete components. However, integrated circuits are being used to an increasing degree, especially in high-performance oscilloscopes, and this trend will doubtless continue and accelerate. Few if any integrated circuits are produced by the major semiconductor manufacturers specifically for oscilloscopes in the way that i.c.s are mass-produced specially for TV sets. The largest oscilloscope manufacturers have their own in-house i.c. facilities, often producing i.c.s in hybrid form, since in scope applications one is always seeking to wring the last ounce of performance out of every circuit. The same consideration is likely to ensure that certain sections of oscillo-

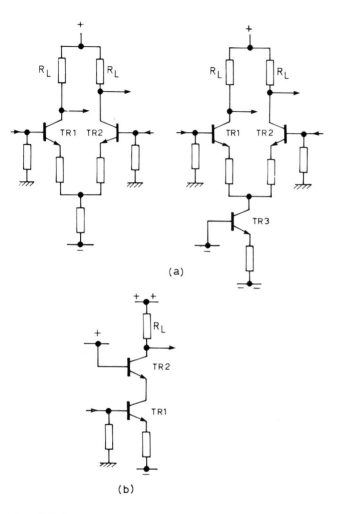

(a)

(b)

Figure 8.2 Basic circuit 'building blocks' commonly used in oscilloscopes: (a) long-tailed pairs, (b) cascode circuit

scopes will continue to be designed using mainly discrete components.

Figure 8.2 shows two of the basic circuit 'building blocks' used in oscilloscopes. The long-tailed pair is widely used in both forms shown, the second being especially common in analogue integrated circuits. It provides balanced push-pull outputs, even if only one input terminal is driven; i.e. it converts from unbalanced to balanced signals. This is an important function, as oscilloscope inputs are usually 'single-ended' or unbalanced, whereas a push-pull or balanced drive is almost invariably applied to the Y (and X) plates. The reason for this is simple. If balanced drive is used, only half the peak-to-peak voltage swing is required at each plate compared to the swing required for the case where only one plate is driven, the other remaining at a constant potential. Thus with balanced drive the supply voltage to the transistors driving the plates can be halved. With only half the voltage across each transistor, the current through it can be doubled without increasing its heat dissipation, which is important in the output stage of a deflection amplifier, as these transistors are invariably run very near the maximum permitted dissipation. With half the supply voltage and twice the current, the load resistor R_L will only be one-quarter of what it would have been for single-ended deflection, resulting in a fourfold increase in bandwidth.

The cascode circuit—*Figure 8.2(b)*— can be seen to consist of a common-emitter stage with a common-base stage as its collector load. This arrangement has two advantages. First, the maximum voltage that can be applied to TR2's collector is equal to the collector-base breakdown voltage V_{cb}, which for high-frequency transistors is often substantially higher than the common-emitter breakdown voltage V_{ce}, enabling a larger output voltage swing to be obtained from the stage. Second, there is inevitably, owing to the construction of a transistor, a capacitance of a few picofarads between its collector and base terminals, denoted C_{cb}. In the cascode circuit, the input capacitance at the base of TR1 is approximately $C_{cb1} + C_{be1}$ (where C_{be1} is the base-emitter capacitance of TR1), since the input impedance at the emitter of grounded-base stage TR2 is very low and there is therefore negligible signal voltage at TR1 collector. If a simple common-emitter stage were used in place of the cascode stage, the input capacitance would appear much larger, as the end of C_{cb} connected to the output would be changing in the opposite sense to the input voltage, by an amount greater than the input voltage swing. In fact, if the stage gain is A, the input capacitance would be approximately $C_{be} + (A + 1)C_{cb}$, the well-known Miller effect. If A is large it would prove difficult to drive the stage satisfactorily, a problem that is avoided by the cascode circuit.

Y deflection amplifier

Oscilloscope designers frequently make use of the advantages of both the long-tailed pair and the cascode, as shown in *Figure 8.3*. Here, the total output capacitance C_t shunting R_L is equal to C_{cb2} plus the load capacitance, several picofarads if this is a deflection plate of a cathode-ray tube. If both transistors have high cut-off frequencies, the −3 dB bandwidth (70.7 per cent response) of the stage is given by $f_{-3\,dB} = 1/2\pi R_L C_t$, showing that for maximum bandwidth both R_L and C_t

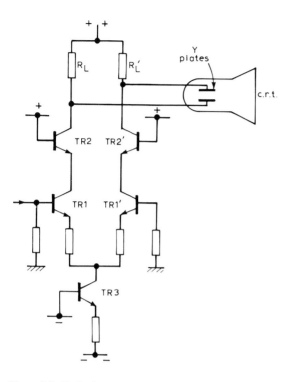

Figure 8.3 Basic deflection-amplifier circuit

should be as small as possible. There is little the oscilloscope designer can do about the plate capacitance of the c.r.t., other than find another tube with the same sensitivity and lower plate capacitance if possible, but TR2 should have both a high collector dissipation rating and a low C_{cb}. Note that if TR2 is changed for another type with twice the dissipation rating, enabling the standing current to be doubled and R_L halved, the bandwidth would be increased even though the C_{cb} of the more powerful transistor were twice that of the original one. This is because C_{cb2} generally constitutes less than 50 per cent of C_t, which will therefore have increased by a much smaller factor than two.

Inductive peaking

A bandwidth greater than the above $f_{-3\,dB}$ can be obtained by the use of inductive peaking circuits to offset the effect of C_t. Note that C_t includes the collector capacitor of the plate-driving transistor, the capacitance of the connecting lead to the plate, and the effective plate capacitance. The last is generally listed by the c.r.t. manufacturer as

113

C_{y1-all}, meaning the capacitance of one Y plate to everything else *except* the other Y plate, and C_{y1-y2}, meaning the capacitance between the Y plates. The effective plate capacitance is $C_{pe} = C_{y1-all} + C_{y1-y2}$ if only one plate is driven, or $C_{pe} = C_{y1-all} + 2C_{y1-y2}$ if, as is usually the case, the two Y plates are driven in antiphase.

Figure 8.4(a) shows a deflection-amplifier output stage using shunt peaking. If we define Q such that $Q = L/R_L{}^2 C_t$, then if L is chosen such that $Q = 0.25$ the pulse response of the stage will show no overshoot, while for $Q = 0.414$ there will be 3.1 per cent overshoot. How-

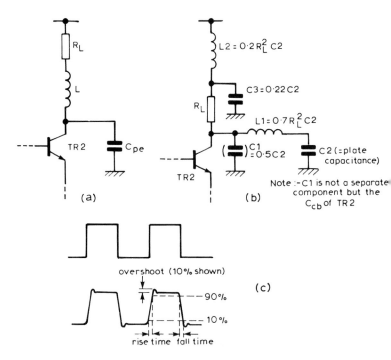

Figure 8.4 Output-stage compensation: (a) simple two-terminal, (b) four-terminal compensation, (c) effect of finite amplifier bandwidth upon an ideal square wave

ever, the rise time will be 71 per cent and 59 per cent respectively of that of the same stage without the inductive peaking. By using a capacitance $C = 0.22C_t$ in parallel with a value of peaking inductance $L = 0.35R_L{}^2 C_t$, the rise time falls to 56.5 per cent of the uncompensated value and the overshoot is only 1 per cent.

The above are examples of 'two-terminal' compensation networks; improved performance at the expense of increased complexity can be obtained by splitting C_t into its components parts. C_{cb} and C_{pe} are

114

compensated separately; the capacitance of the plate connection lead can be included with either of these two to help make up the relative values of capacitance shown in *Figure 8.4(b)*. With this four-terminal peaking circuit, the rise time is only 40 per cent of that of the amplifier without compensation, and overshoot is less than 1 per cent. The improvement in frequency response is much less marked than the reduction in rise time, although if different L and C values are chosen a circuit can be produced with a frequency-response level up to 2.4 times the -3 dB point of the uncompensated amplifier. However, this is of limited use in an oscilloscope as it shows a marked degree of overshoot on fast pulses. Overshoot is illustrated in *Figure 8.4(c)*.

The whole subject of peaking is covered succinctly in Chapter 9 of *Electronic and Radio Engineering* by F. E. Terman, McGraw-Hill, 4th edition, 1955, where an extensive list of further references can be found.

Emitter compensation

With the inductive peaking schemes described above, the improvement in rise time over an uncompensated amplifier is independent of the amplitude of the displayed trace, and is limited to a factor of about 2.5:1 using a four-terminal compensation network. The trend recently has been to abandon inductive peaking of deflection-amplifier output stages in favour of emitter compensation.

This scheme is exemplified in *Figure 8.5*, which shows the circuit of a Y amplifier designed by the author for minimum rise time when using a 3BP1, an insensitive and very outmoded design of c.r.t., but cheap and readily available. Here, the gain of the output amplifier output stage at d.c. and over most of its frequency range is determined by R326, but at higher frequencies C309, 310 tend to bypass R326, resulting in a gain that rises with frequency, compensating for the loading effect of C_t. In fact, the gain of the amplifier transistors is also beginning to fall, with the result that it is not a simple RC load circuit that we are trying to compensate. Consequently, additional components R325, C311 and R308, C314 are included to ensure the smooth roll-off of the frequency response necessary for the faithful reproduction of pulse waveforms.

This type of circuit makes use of the fact that a deflection amplifier is always designed to be able to overscan the available screen display area by up to 100 per cent or more, so that the spot can be deflected way beyond the top or bottom of the graticule. When a very fast rising edge is applied to the Y amplifier, the long-tailed pair TR305, 306 will be overdriven, as their emitters are tied together by C309, 310. The result is that all the available tail current (set by R333; TR307, 308 serve only to introduce the Y shift voltage) is momentarily diverted through, say, TR305 while TR306 is cut off. The load capacitance C_t

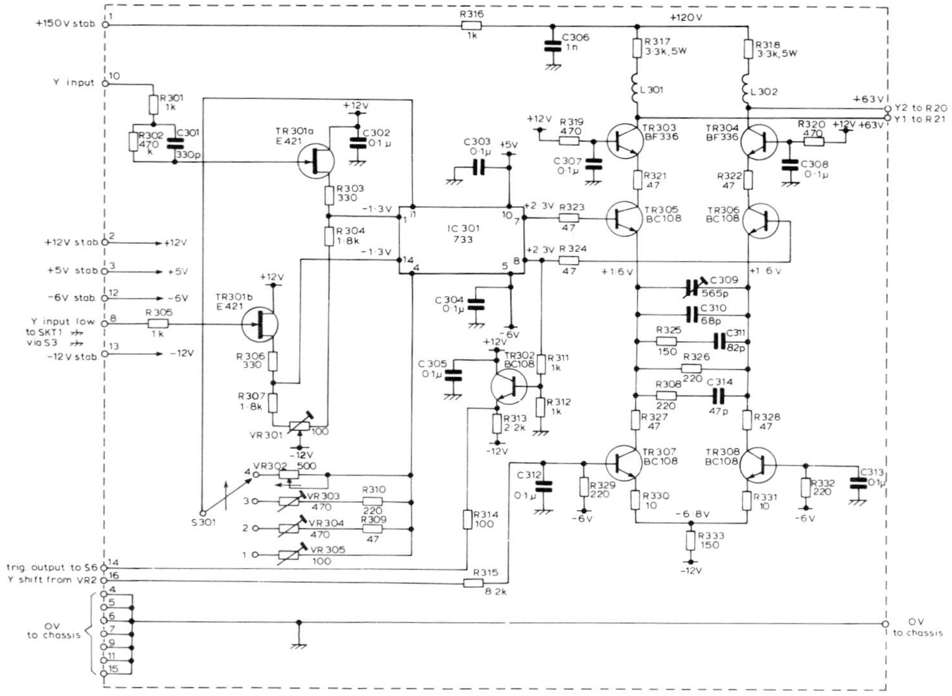

Figure 8.5   Y-deflection amplifier designed by the author for use with c.r.t. type 3BP1

at each collector is therefore charged at the maximum possible rate set by the available tail current. As C_t charges, so do the emitter-compensation capacitors C309 and C310, resulting in the steady-state deflection being reached with minimal overshoot.

This deflection amplifier is said to be 'slew-rate limited' (*Figure 8.6*), as the maximum speed at which the Y-plate voltage can change is determined by C_t and the magnitude of the tail current. Thus, in contrast to inductive peaking, with emitter compensation fast square waves

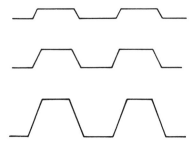

Figure 8.6 Output of slew-rate limited amplifier for three increasing input amplitudes of an ideal square wave

are reproduced more faithfully when they are displayed at small amplitude than when displayed at full screen height. Likewise, the −3 dB bandwidth is greater for small deflections than large; this explains the growing practice of quoting bandwidths at half-screen deflection, which might possibly be reasonable in the case of dual-trace instruments, but is really not fair in a single-channel scope.

Input attenuator

We have dwelt at some length on the Y amplifier because it determines the bandwidth and thus in large measure the usefulness of an oscilloscope. However, the other sections such as the Y input attenuator, trigger and timebase (sweep generator) departments are equally important, so let us complete the description of the Y deflection system by looking at the input attenuator.

The Y amplifier of an oscilloscope normally runs at a fixed value of gain, equal to what it provides on the most sensitive range. For the less sensitive ranges, the input signal is attenuated to bring it down to the same level as on the most sensitive range. Normally, wideband unbalanced variable attenuators are designed with a low, purely resistive characteristic impedance. However, as previously stated, for oscilloscope

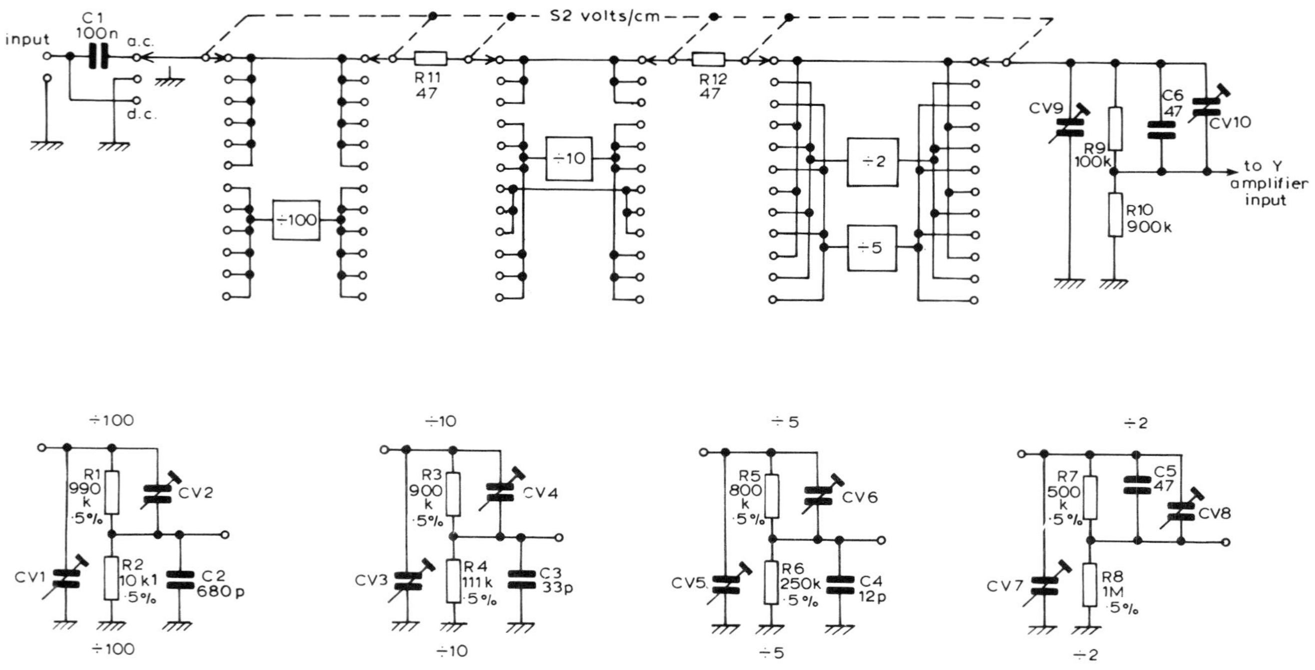

Figure 8.7 Frequency-compensated input attenuator as used in the Scopex 456 (courtesy Scopex Instruments Ltd)

work a high input impedance (especially at low frequencies) is generally required, the standard value being 1 MΩ. At this impedance level the stray capacitance associated with the attenuator resistors, switch, etc. must be taken into account if the attenuation is to remain constant over a bandwidth of even a few megahertz, let alone hundreds of megahertz. This is achieved by absorbing the stray capacitance of the components into larger, deliberately introduced capacitances and then adjusting the latter so that the frequency response is constant.

Figure 8.7 shows a typical input attenuator such as might be found in an oscilloscope of 5 to 10 MHz bandwidth. It is in fact that used in the Scopex 4S6 oscilloscope. It can be seen that each attenuator pad, e.g. the ÷10 position using R3 (900 kΩ) and R4 (effectively 100 kΩ owing to R9 plus R10 in parallel with it) has a capacitive divider CV4 and C3 in parallel. CV4 is adjusted so that its value is one-ninth of C3 + CV9 + the input capacitance of the Y amplifier. Thus the resistive and capacitive division ratios are the same and the attenuation is independent of frequency. CV9 is used to set the input capacitance of the scope on the most sensitive range to a standard value, while CV1, CV3, CV5 and CV7 enable this same input capacitance to be achieved on all the other input ranges. This is important when using a passive divider probe, as described in Chapter 4. R9, C6 and CV10 protect the field-effect transistor forming the first stage of the Y amplifier from damage in the event of a large input at d.c. or low frequency being applied to the oscilloscope when on the most sensitive range, while passing high frequencies largely unattenuated. It is therefore important that large-amplitude signals at high frequencies, e.g. the output of a radio transmitter, should not be applied to an oscilloscope on the more sensitive ranges, as damage may result. It is also worth noting that the input impedance of an oscilloscope is not constant. At d.c. it is 1 MΩ, and virtually 1 MΩ up to a few hundred hertz. Thereafter, it becomes predominantly a capacitive reactance falling with increasing frequency, being typically only 4 kΩ at 1 MHz.

The circuit of *Figure 8.7* is reasonably simple, but it will only perform satisfactorily if the layout is suitable, a comment that applies to the Y amplifier and indeed every section of an oscilloscope. Poor layout or construction in the Y input attenuator can result in partial shunting of the series elements of one pad by the unused components of other ranges. This will result in a non-constant frequency response, which will result in its being impossible to obtain a true square-wave response, except on the most sensitive range where no attenuation is in circuit. Needless to say, the attenuator shown in *Figure 8.7* and incorporated in the 4S6 oscilloscope is designed with intersection screens, to avoid such problems.

Trigger, timebase and X deflection circuitry

Figure 8.8 is the circuit diagram of the trigger-processing circuits, time-base and X deflection amplifier of a dual-trace 15 MHz oscilloscope, manufactured by Gould (formerly Advance Ltd). It is a good example of the tendency noted earlier for modern oscilloscope designs increasingly to incorporate integrated circuits while retaining discrete components for those circuit functions where they are more appropriate. The various sections of the circuit are labelled (e.g. ramp generator, X output amplifier etc.) and detailed operation is described below, as it is typical of modern oscilloscope practice.

The trigger source switches, S502 and S503, connect the required trigger signal via the trigger coupling switches, S504 and S505, to the trigger buffer amplifier formed by TR601 and TR602. S502 selects the differential CH1 signal via R313 and R314 from IC301. S503 selects the equivalent CH2 signal via R363 and R364 from IC351. Where both S502 and S503 are selected, both of the above signals are disconnected and the single-sided input from the EXT TRIG input socket SKC is selected.

When the a.c. coupling switch, S504, is out, the trigger signals are directly coupled-through, but when this switch is in, a.c. coupling is introduced via C603 and C604 (C601 on external). TR601 and TR602 form a differential buffer amplifier with the d.c. balance controlled by the trigger level control, R602. The differential output from this stage is applied to the comparator, IC602, which has positive feedback applied by R623 to form a Schmitt trigger circuit. The change-over switch, S506, reverses the output from TR601 and TR602 to determine the trigger slope.

When both S504 and S505 are 'in' (a.c. and d.c. in for TV mode), the junction of R603 and C610 is connected to the −11 V supply. D601 and D608 are brought into conduction, while D602 and D604 are reverse biased. This diverts the output of the trigger amplifier away from IC602 into TR605, which amplifies the positive tips of the waveform only. TR605 is prevented from saturation by feeding back the peak detected synch pulses via TR607 and TR606 to the emitter of TR605. These pulses are amplified by IC601b and applied via R617 and D603 to the Schmitt trigger, IC602. IC601a is used in conjunction with S504 and S505 to disable the sync separator when a.c. or d.c. is selected.

At the fast timebase sweep speeds, S262a is open and TR603 is cut off. However, at speeds of 100 μs/cm and slower, R608 is connected to +11 V and TR603 is switched on. This effectively grounds C609 to introduce an RC integrating time constant into the sync pulse signal path in the TV mode to separate out the frame trigger.

The square-wave trigger output from IC602 is applied (with d.c. bias of zener diode, D605) as the clock to the D-type TTL flip-flop, IC501a.

A positive-going trigger edge will clock the bistable, driving $\bar{Q}$ low. In the waiting state, $\bar{Q}$ was high (+4.5 V), turning on TR261 via R507 and R262, holding the input (and hence the output) of the operational amplifier, IC261, at 0 V. This timebase amplifier is connected as a direct voltage follower.

When the trigger signal sends $\bar{Q}$ of IC501a low, the timebase clamp transistor, TR261, is turned off. Part of the constant current generated by TR264 flows through the resistor network, R272, to charge C263 at a constant rate. The resultant positive-going linear ramp voltage generated at the input of IC261 is buffered by the amplifier to generate the low-impedance ramp output.

The timebase range switch, S262, selects the tap point on the network, R272, to vary the ramp slope in the 1, 2, 5 sequence over a range of three decades. On all fast sweep ranges TR262 is biased off, but on ramps 0.5 ms/cm and slower S262c connects R263 to +11 V. TR262 is turned on and C264 is effectively connected in parallel with C263 to slow the sweep rate 1000 times.

The constant current in the ramp generator is derived from the current mirror circuit formed by TR263 and TR264. The variable gain control, R269, provides an approximate 3:1 range of variation in this current; R506 provides a preset calibration control on the slow sweep rates, only when S262 is closed.

When the ramp reaches its maximum level, the negative bias introduced by R521 and R519 is overcome and TR503 turns on, driving the reset input of the timebase bistable low. As the bistable switches, $\bar{Q}$ returns high, and TR261 conducts to discharge the timing capacitor(s) and the sweep is complete. However, a hold-off action takes place to inhibit trigger signals during the sweep; this remains for a short period after a sweep to ensure that the ramp potential is fully reset before the next sweep can be triggered. As the ramp goes positive, D506 conducts to charge C502, reverse biasing D503 and turning on TR502. At the end of the sweep when the timebase is reset, $\bar{Q}$ goes low and the D input follows via the action of D508 and R511. The ramp output returns rapidly towards 0 V, but TR502 remains in conduction for a period determined by C502 and R518. Only when TR502 turns off can R516 and D507 take the D input high for the bistable to respond to the next clock input.

TR501 acts in a way similar to TR262 (described above) to introduce additional hold-off time through C501 on the slower half of the timebase ranges.

The brightline facility causes the timebase to free-run in the absence of trigger signals. The square-wave output from the Schmitt trigger, IC602, is coupled via C615 into the peak detector diodes, D606 and D607, to generate a positive-going signal into the negative input of IC601c, driving its output negative. In the absence of such trigger

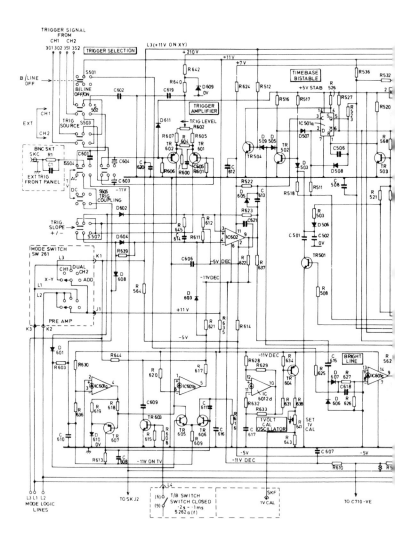

Figure 8.8 X circuits of Gould OS255 oscilloscope, showing timebase generator, trigger circuits and X-deflection amplifier (courtesy Gould Electronics Ltd). The

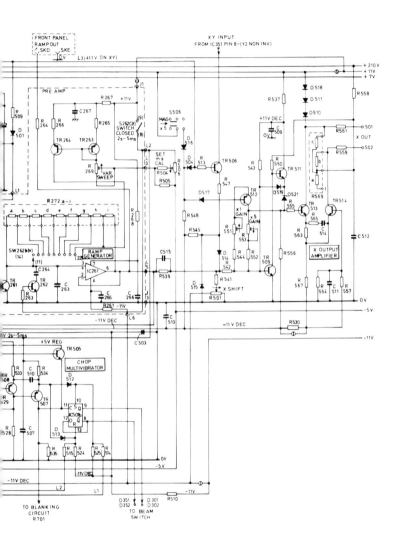

OS255 has been replaced by the OS300 but the circuitry shown above is quite typical of modern practice

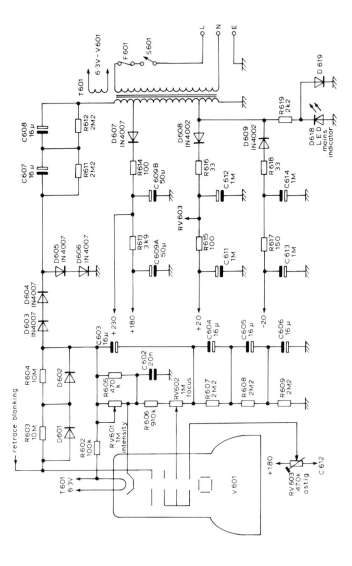

Figure 8.9 Power-supply and c.r.t. circuit of Scopex 456 oscilloscope (courtesy Scopex Instruments Ltd)

signals for a period determined by C618 with R627 and R626, the output of IC601c goes positive. When TR502 turns off at the end of the hold-off period, D509 conducts to turn on TR504, driving the set output low to initiate another sweep. This free-run condition is removed as soon as IC601c detects an output from the Schmitt trigger. It can be inhibited also with a positive bias via R625 if the BRIGHTLINE OFF switch S501 is operated.

The X output amplifier is formed by the shunt feedback stage of TR509/TR511 driving single-sided into the amplifier stage, TR513 and TR514. The collector output of this stage drives the X deflection plates of the c.r.t. The gain introduced by TR509/TR511 is defined in the x 5 magnification mode by the input resistance, R539, and the feedback resistance, R552, with the preset, R553. In this mode the transistor switch, TR512, is biased off. However, in normal x 1 magnification mode, S507 is open and the current in R548 turns on TR512, introducing R544 with preset R551 as additional feedback to reduce the gain of the amplifier accordingly.

The X shift control, R501, introduces an additional bias input via R541 into the input of the shunt feedback amplifier.

Power supply and c.r.t. circuitry

Figure 8.9 is the circuit diagram of the c.r.t. and power supplies section of a straight-forward oscilloscope, the Scopex 4S6 mentioned earlier.

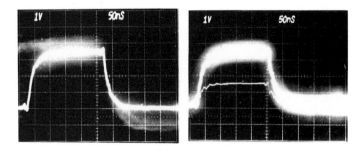

Figure 8.10 The self-limiting effect of the microchannel plate compresses the brightness range – see *Figure 7.11*. Left – A pulse train on a TEK 7904A doesn't reveal the low-level glitch occurring every ten-thousandth pulse. (The TEK 7904A was previously the world's fastest-writing-rate scope.) Right – The same pulse train viewed directly on the 7104/R7103, with one-thousand times the brightness of conventional scopes. The researcher can now analyse the pulse with the naked eye and take pictures with ease (courtesy Tektronix UK Ltd)

125

All the supplies are derived from a mains transformer with an untapped primary, providing operation from 210 to 250 V a.c., 48–60 Hz. The 6.3 V secondary winding that supplies the c.r.t. heater is insulated to withstand the full −1.4 kV e.h.t. voltage applied to the c.r.t. cathode/grid circuit. All the d.c. supplies are derived from a single-tapped secondary winding; as is usually the case in inexpensive scopes, they are not stabilised. This will cause the deflection sensitivity of the c.r.t. to vary with mains voltage, but the design of the Y amplifier is such that its gain varies with mains voltage in the inverse sense, maintaining the overall gain sensibly constant at the calibrated value.

Intensity, focus and astigmatism controls are provided, the first two being mounted on the front panel. However, once set up during production test, the astigmatism control will need re-adjustment rarely if ever, so this control is a preset potentiometer mounted internally.

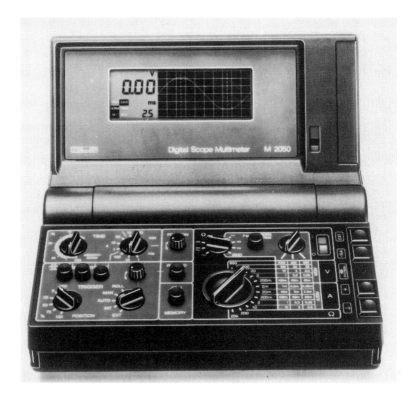

Figure 8.11 The M 2050 Digital Scope Multimeter provides 500K samples/sec digital storage and uses a liquid crystal display. The hinged case closes like a clam shell to protect the controls and display when not in use (courtesy British Brown-Boveri Ltd)

Appendix 1

Cathode-ray tube phosphor data

Human eye response

An important factor in selecting a phosphor is the colour or radiant energy distribution of the light output. The human eye responds in varying degrees to light wavelength from deep red to violet. The human eye is most sensitive to the yellow-green region; however, its responsiveness diminishes on either side in the orange-yellow area and the blue-violet region. The eye is not very receptive to deep blue or red.

If the quantity of light falling on the eye is doubled, the brightness 'seen' by the eye does not double. The brightness of a colour tone as seen is approximately proportional to the log of energy of the stimulus.

The term *luminance* is the photometric equivalent of brightness. It is based on measurements made with a sensor having a spectral sensitivity curve corrected to that of the average human eye. The SI (international metric standard) units for luminance are candelas per square metre, but footlamberts are still used extensively in the US; 1 footlambert = 0.2919 candela/m². The term luminance implies that data has been measured or corrected to incorporate the CIE standard eye response curve for the human eye. CIE is an abbreviation for Commission Internationale de L'Eclairage (International Commission on Illumination). The luminance graphs and tables are therefore useful only when the phosphor is being viewed.

Phosphor protection

When a phosphor is excited by an electron beam with an excessively high current density, a permanent loss of phosphor efficiency may occur. The light output of the damaged phosphor will be reduced, and in extreme cases complete destruction of the phosphor may result. Darkening or burning occurs when the heat developed by electron bombardment cannot be dissipated rapidly enough by the phosphor.

The two most important and controllable factors affecting the occurrence of burning are beam-current density (controllable with the intensity, focus, and astigmatism controls) and the length of time the beam excites a given section of the phosphor (controllable with the time/div control). Of the total energy from the beam, 90 per cent is converted to heat and 10 per cent to light. A phosphor must radiate the light and dissipate the heat, or like any other substance it will burn. Remember, burning is a function of intensity and time. Keeping the intensity down or the time short will save the screen.

Photographic writing rate

Photographic writing rate is a measure of the scope/camera/film's capability to record high speed signals.

Recording high speed signals on film is dependent on at least three factors: the oscilloscope used, film characteristics, and the camera. For maximum writing rate capability, the objective is to get as much light energy to the film surface as possible. Since each component affects photographic writing rate, the selection for top performance is important. The phosphor offering the highest photographic writing rate is BE (P11). A c.r.t. with this phosphor is therefore usually specified for an oscilloscope which is required to record photographically very fast single events, which leave too feint a trace to be observed visually. However, a microchannel plate c.r.t. (*Figure 7.11*) enables one to see clearly single shot events at the full bandwidth of the oscilloscope. For this reason, GH (P31) phosphor is standard on MCP c.r.t.s.

Note The information in this appendix is reproduced by courtesy of Tektronix UK Ltd.

Comparative CRT phosphor data

Phosphor WTDS	JEDEC	Fluorescence and phosphorescence	Relative luminance[2]	Relative photographic writing speed[3]	Decay	Relative burn resistance	Comments
GJ	P1	Yellowish-green	50%	20%	Medium	Medium	Replaced by GH (P31) in most applications
WW	P4	White	50%	40%	Medium-short	Medium-high	Television displays
GM	P7	Blue[5]	35%	75%	Long	Medium	Long decay, double-layer screen
BE	P11	Blue	15%	100%	Medium-short	Medium	For photographic applications
GH	P31	Green	100%	50%	Medium-short	High	General purposes, brightest available phosphor
GR	P39	Yellowish-green	27%	NA[4]	Long	Medium	Low refresh rate displays
GY	P43	Yellowish-green	40%	NA[4]	Medium	Very high	High current density phosphor
GX	P44	Yellowish-green	68%	NA[4]	Medium	High	Bistable storage
WB	P45	White	32%	NA[4]	Medium	Very high	Monochrome TV displays

[1] Tektronix is adopting the Worldwide Phosphor Type Designation System (WTDS) as a replacement for the older JEDEC 'P' number system reference.   The chart lists the comparable WTDS designations for the most common 'P' numbers.

[2] Measured with Tektronix J16 Photometer and J6523 Luminance Probe which incorporates a CIE standard eye filter. Representative of 10 kV aluminized screens. GH (P31) as reference.

[3] BE (P11) as reference with Polaroid 612 or 106 film. Representative of 10 kV aluminized screens.

[4] Not available.

[5] Yellowish-green Phosphorescence.

BE (P11) phosphor has a different spectral output than GH (P31) phosphor standard and more closely matches the sensitivity spectrum of silver halide film types. While photographic writing speed is approximately two times the GH (P31) rate, the visual output luminance is approximately 15% of GH (P31) phosphor standard, using Polaroid Film Type 107, 3,000 ASA w/out film fogging.

Appendix 2

Oscilloscope manufacturers

The following list gives the names, addresses and sales office telephone numbers of most of the manufacturers of oscilloscopes whose products are readily available in the UK. The information is believed to be correct at the date of publication but no responsibility can be taken for errors and omissions; in particular various foreign makes of oscilloscope may be available but not listed. For foreign manufacturers, UK sales office details are given, or a reference is given to a UK agent or distributor, not necessarily the main agent. Manufacturers of related instruments (e.g. panoramic receivers, spectrum, network or logic analysers) are not listed unless they also produce oscilloscopes.

Tel = telephone, Tx = telex.

Advance Bryans Instruments, This is the name of the company previously known as House of Instruments. 14–16 Wates Way, Mitcham, Surrey CR4 4HR. Tel 01 640 5624. Agent for Trio, Soar Corporation, Vuko.

Beckman Industrial, Electronic Technologies Division, UK Sales and Marketing Organisation, Temple House, 43–48 New Street, Birmingham B2 4LJ. Tel 021 339576. Tx 336659.

B&K Precision. see Dynascan Corporation.

Bridage and Scopex, 117 Knowle Road, Mirfield, West Yorks WF14 9RJ. Tel 0924 498703.

British Brown-Boveri Ltd. Normelec Division, Grovelands House, Longford Road, Exhall, Coventry CV7 9ND. Tel (0203) 364021. Tx 312114. UK office representing Brown-Boveri Corporation.

Crotech Instruments Ltd, 2 Stephenson Road, St. Ives, Huntingdon, Cambs PE17 4WJ. Tel 0480 301818.

Dynascan Corporation, International Sales, 6460 West Cortland Street, Chicago, Illinois 60635, USA.

Electronic Brokers Ltd, 140–146 Camden Street, London NW1 9PB. Tel 01 267 7070. Tx 298694. Agents for Grundig, Hameg, Philips, Thander.

Enertec Instrumentation Ltd, see Solartron Instruments.

Farnell Instruments Ltd, Sandbeck Way, Wetherby, West Yorks LS22 4DH. Tel 0937 61961. Tx 557294.

Fieldtech (Heathrow) Ltd, Huntavia House, 420 Bath Road, Longford, Middlesex. UB7 0LL. Tel 01 897 6446. Tx 23734. Agent for Meguro.

Gould Electronics Ltd, Roebuck Road, Hainault, Essex IG6 3UE. Tel 01 500 1000. Tx 263785.

Grundig AG, see Electronic Brokers Ltd.

Hameg Ltd, 74–78 Collingdon Street, Luton, Beds LU1 1RX. Tel 0582 413174. Tx 825484.

Hewlett Packard Ltd, King Street Lane, Winnersh, Wokingham, Berks RG11 5AR. Tel 0734 784774. Tx 847178/9.

Hitachi Denshi (UK) Ltd, 13-14 Garrick Industrial Centre, Garrick Road, Hendon, London NW9 9AP. Tel 01 202 4311. Tx 27449.

House of Instruments (Anglia) Ltd, see Advance Bryans Instruments Ltd

ITT Instruments, 346 Edinburgh Avenue, Slough. Berks SL1 4TU. Tel 0753 824131. Tx 849808. Agent for Metrix.

Iwatsu Electric Co Ltd, see STC Instrument Services.

Kikusui Electronics Corporation, see Telonic Instruments Ltd.

Lawtronics Ltd, 139/141 High Street, Edenbridge, Kent TN8 5AX. Tel 0732 865191. Oscilloscope manufacturer and agent for NLS, Hameg, Kikusui, Crotech.

Leader, see Thandar Electronics Ltd.

Le Croy Research Systems Ltd, Elms Court, Botley, Oxford OX2 9LP. Tel 0865 727275.

Meguro Electronics Corporation, see Fieldtech (Heathrow) Ltd.

Metrix, division of ITT Composants et Instruments, see ITT instruments.

National Panasonic, see Panasonic Industrial UK Ltd.

Nicolet Instruments Ltd, Budbrooke Road, Warwick. Tel 0926 494111. Tx 311135.

NLS Inc, see Lawtronics Ltd.

Panasonic Industrial UK Ltd, 280/290 Bath Road, Slough, Berks SL1 6JG. Tel 0753 73181. Contact UK agent Wavetek for sale enquiries.

Philips, see Pye Unicam Ltd.

Pye Unicam Ltd, York Street, Cambridge CB1 2PX. Tel 0223 358866. Tx 817331. Agent for Philips.

Salford Electrical Instruments Ltd, Times Mill, Heywood, Lancs OL10 4NE. Tel 0706 67501. Tx 635106. Manufacturers of intrinsically safe (IS Approval, Group 1) oscilloscopes.

Schlumberger, see Solartron Instruments.

Siemens AG, see Siemens Ltd.

Siemens Ltd, Siemens House, Windmill Road, Sunbury on Thames, Middlesex. Tel 09327 85691. Tx 8951091. Agent for Siemens AG.

Soar Corporation, see Advance Bryans Instruments.

Solartron Instruments, Victoria Road, Farnborough, Hants GU14 7PW. Tel 0252 544433. Tx 858245. A Division of Schlumberger Electronics (UK) Ltd.

STC Instrument Services Ltd, Dewar House, Central Road, Harlow, Essex CM20 2TA. Tel 0279 29522. Tx 817202. Agent for Iwatsu.

Tektronix UK Ltd, PO Box 69, 36/38 Coldharbour Lane, Harpenden, Herts AL5 4UP. Tel 05827 63141. Tx 25559.

Telonic Instruments Ltd, Boyn Valley Road, Maidenhead, Berks SL6 4EG. Tel 0628 73933. Tx 849131. Agent for Kikusui.

Thandar Electronics Ltd, London Road, St. Ives, Huntingdon, Cambs PE17 4HJ. Tel 0480 64646. Tx 32250. Oscilloscope manufacturer and agent for Leader.

Trio Kenwood Corporation, see Advance Bryans Instruments.

Vuko Elektronische Geräte GmbH, see Advance Bryans Instruments.

Wavetek Electronics Ltd, Tag Lane, Hatch End, Reading, Berks. RG10 9LT. Tel 073 522 4121. Tx 849301 Agent for Panasonic.

Index

132